Lecture Notes in Electrical Engineering

Volume 281

For further volumes:
http://www.springer.com/series/7818

About this Series

"Lecture Notes in Electrical Engineering (LNEE)" is a book series which reports the latest research and developments in Electrical Engineering, namely:

- Communication, Networks, and Information Theory
- Computer Engineering
- Signal, Image, Speech and Information Processing
- Circuits and Systems
- Bioengineering

LNEE publishes authored monographs and contributed volumes which present cutting edge research information as well as new perspectives on classical fields, while maintaining Springer's high standards of academic excellence. Also considered for publication are lecture materials, proceedings, and other related materials of exceptionally high quality and interest. The subject matter should be original and timely, reporting the latest research and developments in all areas of electrical engineering.

The audience for the books in LNEE consists of advanced level students, researchers, and industry professionals working at the forefront of their fields. Much like Springer's other Lecture Notes series, LNEE will be distributed through Springer's print and electronic publishing channels.

Koen Langendoen · Wen Hu
Federico Ferrari · Marco Zimmerling
Luca Mottola
Editors

Real-World Wireless Sensor Networks

Proceedings of the 5th International Workshop, REALWSN 2013, Como (Italy), September 19–20, 2013

Editors
Koen Langendoen
Delft University of Technology
Delft
The Netherlands

Wen Hu
CSIRO ICT Centre
Queensland Centre for Advanced Technologies
Pullenvale
Australia

Federico Ferrari
Marco Zimmerling
Computer Engineering and Networks Laboratory
ETH Zurich
Zurich
Switzerland

Luca Mottola
Dipartimento di Elettronica, Informazione e Bioingegneria
Politecnico di Milano
Milan
Italy

ISSN 1876-1100 ISSN 1876-1119 (electronic)
ISBN 978-3-319-03070-8 ISBN 978-3-319-03071-5 (eBook)
DOI 10.1007/978-3-319-03071-5
Springer Cham Heidelberg New York Dordrecht London

Library of Congress Control Number: 2013954846

Printed on acid-free paper

Springer is part of Springer Science+Business Media (www.springer.com)

Preface

It is our great pleasure to welcome you to REALWSN 2013, the 5th Workshop on Real-World Wireless Sensor Networks, with a focus of bringing together researchers and practitioners to discuss state-of-the-art and best practices in sensor networks.

This year we received 32 submissions, originating from over 18 different countries around the globe. Apart from a few submissions violating the double-blind requirement, all papers received at least three reviews and engaged in an online discussion phase. Finally, 10 full papers and 6 short papers were selected for this year's program. As the sensor network field is maturing the program includes a wide variety of topics, ranging from RF front ends and directional antennas, via MAC protocols and routing, to applications studies and real-world deployments. In addition to the technical program, the workshop features a poster and demo session with nine entries, making for an exciting event all-in-all.

We thank the members of the technical program committee, the poster and demo chairs (Thiemo Voigt and Silvia Santini), the publication chairs (Federico Ferrari and Marco Zimmerling), and the local organizers (Mikhail Afanasov and Alessandro Sivieri) for their contributions to the organization of the workshop. Above all, we would like to thank Luca Mottola, the general chair, for keeping us on track and making REALWSN 2013 a reality.

September 2013

Wen Hu
Koen Langendoen

Organization

REALWSN 2013 was organized by Politecnico di Milano, Dipartimento di Elettronica, Informazione, e Bioingegneria.

General Chair

Luca Mottola	Politecnico di Milano, Italy and SICS Swedish ICT, Sweden

Program Co-Chairs

Wen Hu	CSIRO, Australia
Koen Langendoen	TU Delft, The Netherlands

Poster and Demo Chairs

Silvia Santini	TU Darmstadt, Germany
Thiemo Voigt	Uppsala University and SICS Swedish ICT, Sweden

Program Committee

Nirupama Bulusu	Portland State University, USA
Per Gunningberg	Uppsala University, Sweden
Chamath Keppitiyagama	University of Colombo, Sri Lanka
Gian Pietro Picco	University of Trento, Italy
Utz Roedig	Lancaster University, UK
Christian Rohner	Uppsala University, Sweden
Kay Römer	TU Graz, Austria
Jochen Schiller	FU Berlin, Germany
Cormac Sreenan	University College Cork, Ireland
Tim Wark	CSIRO, Australia

Neal Patwari University of Utah, USA
Omprakash Gnawali University of Houston, USA
Yu (Jason) Gu Singapore University of Technology and Design
Olga Saukh ETH Zurich, Switzerland
Niki Trigoni University of Oxford, UK
Marco Zúñiga TU Delft, The Netherlands
Prasant Misra SICS Swedish ICT, Sweden
Chiara Petrioli University of Rome 'La Sapienza', Italy
Philipp Sommer CSIRO, Australia
Gianluca Dini University of Pisa, Italy
Wang Jiliang Hong Kong University of Science and Technology
Geoffrey Challen University at Buffalo, USA
Marcus Chang Johns Hopkins University, USA

Publication Chairs

Federico Ferrari ETH Zurich, Switzerland
Marco Zimmerling ETH Zurich, Switzerland

Publicity Chair

Mikhail Afanasov Politecnico di Milano, Italy

Web Chair

Alessandro Sivieri Politecnico di Milano, Italy

Contents

Part I
Applications

Snowcloud: A Complete Data Gathering System for Snow Hydrology Research

Christian Skalka and Jeffrey Frolik

Abstract Snowcloud is a data gathering system for snow hydrology field research campaigns conducted in harsh climates and remote areas. The system combines distributed wireless sensor network technology and computational techniques to provide data to researchers at lower cost and higher spatial resolution than ground-based systems using traditional "monolithic" technologies. Supporting the work of a variety of collaborators, Snowcloud has seen multiple Winter deployments in settings ranging from high desert to arctic, resulting in over a dozen node-years of practical experience. In this chapter, we discuss both the system design and deployment experiences.

1 Introduction

The ability to characterize snowpack state, as well as snowmelt, is broadly important for understanding hydrological and ecological processes and incorporating those processes in agricultural, ecological, etc. models [11]. Snowmelt is the primary source of water in many mountainous regions of the world and as a result is a critical necessity for about 16 % of the world's population [16]. Current climate model simulations show that snow processes are not stationary [2] and observations show snowpack has declined across much of the US in recent decades [3]. Despite the importance of data gathering in this realm, there exist major gaps in observing snowmelt and runoff [14, 20], even in relatively well-instrumented regions of the US. Current observations are relatively sparse and correlations among point measurements and model estimates can vary significantly [15]. Improved snow observations are thus desperately

C. Skalka (✉) · J. Frolik
University of Vermont, Burlington, VT, US
e-mail: ceskalka@cems.uvm.edu; skalka@cs.uvm.edu

J. Frolik
e-mail: jfrolik@cems.uvm.edu

K. Langendoen et al. (eds.), *Real-World Wireless Sensor Networks*,
Lecture Notes in Electrical Engineering 281, DOI: 10.1007/978-3-319-03071-5_1,

needed to provide objective measures for verification of hydrologic model forecasts [18] and to better streamflow predictions through updating the modeled snow water equivalence (SWE) [5].

Wireless sensor networks can address this need, especially for ground-truth data gathering. WSNs have significant advantages over existing methods in terms of combined temporal and spatial resolution, deployment flexibility and low environmental impact, and low cost. Snow courses are accurate, but invasive, human-resource intensive, and usually have poor temporal resolution. Traditional ground-based automated sensors such as SNOTEL sites have good temporal resolution, but are limited in terms of spatial resolution due to the high cost of deployment and maintenance. Finally, both manual surveying and SNOTEL sites are ill-suited to forested areas and highly variable topologies, settings in which spatial and temporal variability of snowpacks need to be better understood.

Our system, Snowcloud, leverages the advantages of WSNs for snow hydrology research. It was specifically developed as an instrument for short- and medium-term field research campaigns in remote locations, that could be used by a variety of researchers, and easily re-tasked to a diverse range of studies. Snowcloud was thus designed to be low-cost and within the budget constraints of academic researchers, to be modular for ease of shipping/transport to and assembly at remote locations, and to not be dependent on any existing infrastructure for data collecting. Furthermore, Snowcloud is a *complete* system, comprising data production, collection, and presentation. Our online presentation of data also anticipates public use.

Other projects have previously leveraged WSN technology to study cold-lands processes. Embedded wireless sensing has been used to study glacial movement [6] and permafrost [7]. In addition, WSNs have been proposed to better under snow in terms of structure [10] and conditions leading to avalanches [1, 8, 17]. Most closely related to our work is an extensive, long-term network deployed at the southern sierra critical zone observatory (CZO) [9]. As a complement to the extensive CZO site instrumentation, this wireless sensor network consists of 23 nodes each with an extensive suite of science-grade instrumentation along with additional 34 nodes to ensure network connectivity. In comparison to alternative methods (e.g., wired data loggers), the CZO deployment provides the ability to collect data nearly continuously and present it in near real-time from across the 1 km^2 study site. But in contrast to our work, this is a longer-term, larger-scale, higher cost project, designed for a very specific purpose. Snowcloud is intended to be a smaller, more affordable tool for use in a broad range of studies.

Herein, we present the Snowcloud system in the context of the life cycle of data from sampling to storage and presentation (Fig. 1). We highlight the technical details that support our stated aims. In Sect. 2, we describe the network hardware and software platforms, and how data is sampled and formatted. In Sect. 3, we discuss solutions employed to collect and report data. In Sect. 4, we present an oft overlooked aspect of WSN, specifically the processing of data and its presentation to end users via a publicly available database. We discuss several Snowcloud deployments to date in Sect. 5 along with some key technical and practical experiences. We conclude by discussing future work related to algorithms and sensors.

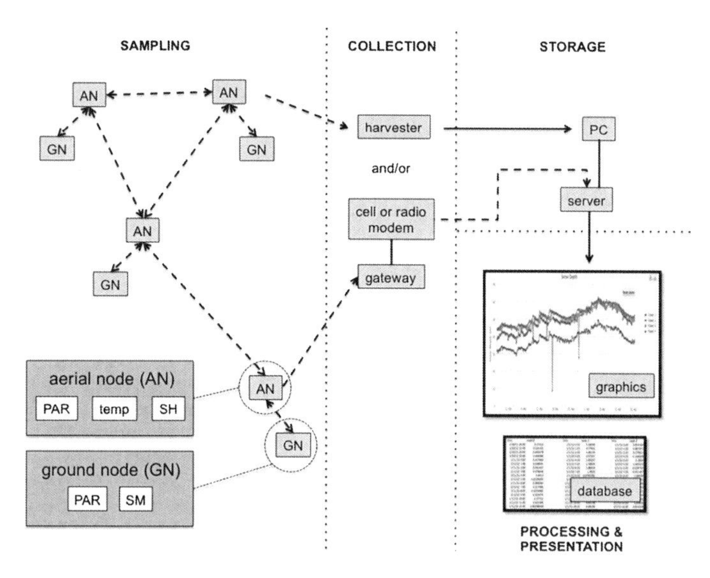

Fig. 1 Snowcloud system components and data life cycle

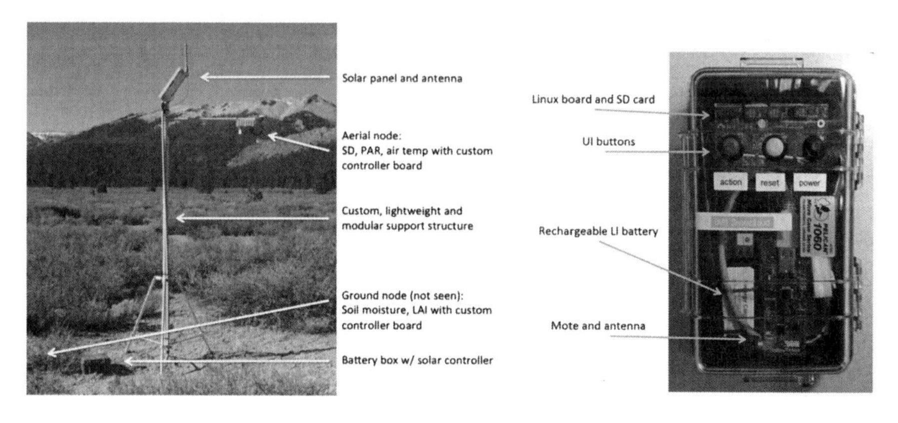

Fig. 2 Snowcloud tower (*L*) and Harvester device (*R*)

2 Sampling Data

The Snowcloud network consists of multiple towers (Fig. 2), each hosting one or two wireless sensor devices (i.e., nodes) that collect the pertinent data. Nodes are deployed above the snowpack (i.e., aerial nodes) and sometimes below the snowpack (i.e., ground nodes) depending on the science objectives. Nodes communicate via a TinyOS mesh network. Depending on the sensor suite and battery requirements, a completed single tower ranges in cost from \$500 to \$1,000. We detail these various aspects of the platform in the following paragraphs.

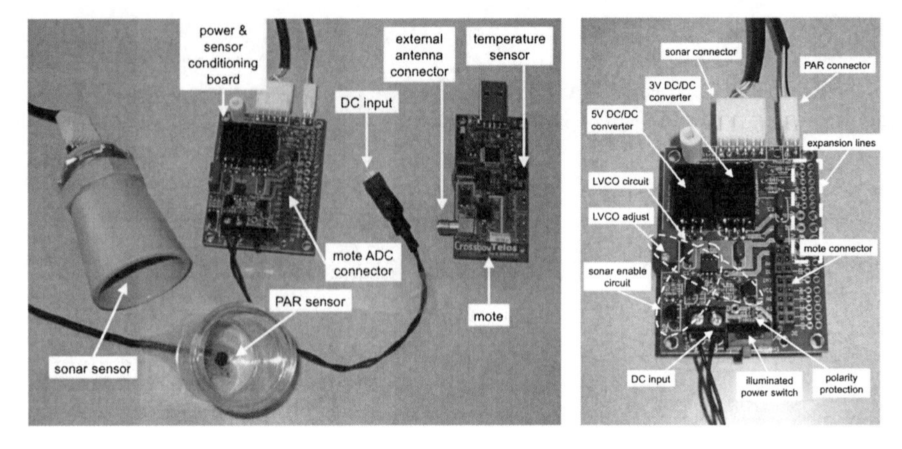

Fig. 3 Snowcloud sensors and electronics (*L*) and control board features (*R*)

Computation, Timing, and Communications The nerve center of each tower is a MEMSIC TelosB mote [12], pictured in Fig. 3, running the TinyOS2 operating system. We have developed a suite of programs for sensor control, including power cycling and sample rates, and on-mote datalogging and reporting. Regardless of the remote data collection method used, each tower logs sampled data in non-volatile flash memory on the mote for backup. The on-board clock is used to time sampling epochs, and each sample is timestamped with node-local time.

Although network time synchronization protocols such as FTSP are available in TinyOS, using node-local time without network synchronization has generally been adequate for existing deployments. There has been very little node power cycling, and deployments are typically serviced and restarted within 8 months—TelosB clock drift during such a period is tolerable in this application. Also, protocols such as FTSP are intended for much more precise time synchronization than we need. The benefit of ignoring time synchronization is simplification of code development, which is non-trivial since node programs are already quite complex and difficult to debug. Also, new gateway technology as discussed in Sect. 3 will provide the network with a battery backed real-time clock. However, network time synchronization would certainly provide a more robust system and allow nodes to periodically operate in low-power mode, so we intend to include time synchronization in future iterations of the system.

Custom Control Board We have developed a custom control board (Fig. 3) with a number of hardware features useful in this application space. The control board includes basic features such as voltage regulators for the mote and sensors and breakouts for the mote ADC pin array. It also contains a switch allowing the mote to power cycle sensors, supporting an energy-efficient power regime defined at the software level—in short, active sensors are powered off when they are not sampling. The board includes a low voltage cutoff (LVCO) circuit to protect draw-down of batteries in case solar recharging is interrupted for extended periods, for example during

Winter storm cycles, or due to solar panel snow loading. A voltage sensor is also incorporated to monitor solar panel/battery voltage.

Sensor Systems and Sampling Regime The Snowcloud system can be configured to support a variety of sensors. The current "standard" configuration for the aerial node includes an ultrasound sensor, and air temperature sensor, a photosynthetically active radiation (PAR) sensor, and a system voltage sensor. The low-cost ultrasound sensor is a ruggedized Maxbotix sensor with a 15 cm to 4 m range, which produces a voltage proportional to the round-trip time of flight. These sensors are pictured in situ in Fig. 2, and in detail in Fig. 3. We have also implemented ground nodes for measuring soil moisture and ground level PAR, the latter being useful for ascertaining bush leaf-area-index (LAI). Due to the short link lengths, ground nodes have no difficulty, using just a patch antenna, communicating through the snowpack with aerial nodes to disseminate collected data.

Sampling intervals are determined by user requirements and expected solar exposure and power availability. As snowpack evolution is a slow process, typical sampling intervals utilized for Snowcloud deployments are either 1 or 3 h. The ultrasound sensor is powered off by software when not sampling. Multiple samples are taken in each sampling cycle, aka *epoch*, typically 12 ultrasound readings, 5 PAR readings, and 3 temperature readings. Only median values for each sensor type are stored in mote flash memory to conserve log space.

Power System Snowcloud towers are powered by a combination of 12 V lead acid batteries and a 12 W photovoltaic panel. This is a popular solution and many related products are available, including solar controllers. Although lead acid batteries are heavy, robustness to cold temperatures and a wide recharge range make them our preferred choice. The TelosB platform has a 20–30 milliamp draw on average, which is easily powered by the solar panel in full solar exposure. However, adequate battery power is required at night, during extended storm cycles, and at the depths of Winter in arctic deployments. Deployed battery capacity has varied from 12 to 55 amp h depending on deployment conditions. In all cases, the control board's LVCO prevents battery draw down below a 10 to 11 V adjustable threshold. The LVCO circuit is to prevent deep battery discharge as most solar controllers will not recharge batteries drawn down below 9 V. If a node is shut off by the LVCO, it will automatically restart when battery charge comes back above the threshold.

Support Structure and Enclosures As seen in Fig. 2, the Snowcloud support structure consists of a vertical mast from which the aerial node is cantilevered. At the top of the mast is the solar panel and communication antenna. The standard tower for deployment in areas with high annual snowfall provides approximately 2.5 m of clearance between the ultrasound sensor and the ground. The mast is readily assembled from 1 m segments of aluminum thereby allowing tower height to be readily increased or decreased, and easily packed. The structure has been tested in Solidworks® and is designed to withstand winds up to 100 mph. The most challenging aspect of the structure design is the anchoring mechanism as the ground at our deployment sites has ranged from granite to sand to bog. We have used both a plate anchor that is affixed to the ground, and a tripod base combined with a buried ballast (e.g., a plastic bucket filled with rock). The latter approach is easily installed and results in more stable structure.

For electronics enclosures, we have used Pelican® cases of various sizes. Especially for batteries, these are relatively cheap, adaptable solutions, and can be easily drilled to accommodate wiring pass-throughs.

3 Collecting and Storing Data

We now consider how we collect and report data. By "collect" we mean how we manage voltage samples as data once they are registered on mote ADC ports. By "report" we mean how we communicate that data to a permanent storage device, i.e., a database on a lab-accessible file server. The interpretation and visualization of pure voltage data is treated as a separate matter in Sect. 4.

Data Storage Layers and Redundancy In the Snowcloud system, data is potentially stored at three layers: permanent storage, a data collection device, and non-volatile flash memory on the nodes themselves. Each node's flash memory space is adequate to store a year of data for sensors with hourly sampling rates. In our experience, this storage mechanism is highly robust and can always be relied upon when all else fails. The use of data collection devices, described in detail below, provides more reliability, convenience, and near-real-time data reporting, and also interesting automated systems control opportunities that we envision as future work.

Data Harvester and Pull Protocol We have developed and implemented a hand-held *Harvester* device to serve as the primary data collection device for areas without cellular coverage. Users transport the device to and from the site, and collect data while in network proximity by issuing a command from a simple push-button interface. The device is waterproof for use in snow, and has a rechargeable lithium-ion battery. The Harvester leverages TinyOS ad-hoc mesh networking, so that a communication link only needs to be established between it and one arbitrary tower in a connected network.

Device status during use is provided by built-in LEDs on processor boards, while input is provided by external buttons wired to the user and reset buttons on a TelosB mote inside the Harvester. This mote establishes a network connection with the Snowcloud deployment and issues requests for data. Reported data is relayed via USB to a Technologic Systems TS7260 board with 12GB of flash memory, where it is available for subsequent download in the lab e.g. via ethernet. Harvester operation is based on a custom-designed pull protocol layered over the TinyOS *Dissemination* protocol and *collection tree protocol* (CTP). The protocol provides a "push button" user experience, where a single button push initiates collection of all data within the network. The protocol will not interfere with normal network operation, i.e., sampling. It is scalable to arbitrary network size, and is robust to node failure during reporting. Otherwise, the protocol does not provide integrity or reliability guarantees beyond those provided by CTP. We impose a data reporting flow control for the connection between the mote and the TS7260, since in testing we encountered data loss without it. Total collection times vary depending on number of nodes, length of deployment, and sampling rates, but after a few months of deployment pulling data tends to take between 10 and 60 min.

Most interestingly, when a Harvester device is introduced to the network by the user, it becomes a CTP "root node" to receive data—but although it is not well-documented, our field experience has revealed that CTP does not support a network with *zero* root nodes, which is the case when the Harvester is removed. Rather, CTP that has been running for about a day or more will no longer accept roots and report data. Thus, the Harvester pull protocol uses Dissemination to stop and restart CTP at the removal and reintroduction of the device by the user.

Data Gateway and Push Protocol We are currently developing and testing a *Gateway* device, that will receive data from the sensor network and report it in near-real-time over the Internet. This device is essentially the same hardware platform as the Harvester, coupled with a cellular modem, a battery backed real time clock, and an external power supply. The Gateway receives and stores sensor network data in a local mySQL database. This provides a second layer of data redundancy in the system—in the absence of cellular connectivity, data can be manually retrieved from the gateway device, e.g., by pulling the SD card. In the presence of cellular connectivity, a program periodically runs on the gateway and reports new data over the GSM cellular modem to the Internet via FTP. Periods are application dependent.

The Gateway currently uses CTP and a push protocol in the network. Nodes report samples to it as they are taken. The Gateway timestamps samples as they are received. Note that this protocol is more robust to node failure: in particular, if the network pull protocol is used and a node stops and restarts due to battery charge and LVCO operation, the "restart time" of the node cannot be known and subsequent node local timestamps cannot be correlated with real time. In contrast, the Gateway can always assign real timestamps when samples are immediately pushed by sensor nodes. And in the event of Gateway stop and restart, a Harvester-type pull protocol can be automatically run on restart to retrieve missed data.

4 Processing and Presenting Data

Data Pipeline Whether a Harvester or Gateway is used to collect data from the network, it is initially available in permanent storage in flatfiles. Each entry records the mote ID, the sensor type, data represented in ADC counts, and a sample timestamp. Node local timestamps are automatically converted to real timestamps given the known node start time. This data is easily parsed and entered into a relational database. Once in the database, data processing scripts are applied to obtain physical interpretations of sensor voltages as described below, e.g., ultrasound and temperature sensor samples are combined to obtain snow depth readings. It is then available to users via online web interfaces.

Data Processing and Interpretation The final product of the Snowcloud system is processed sensor data. An example of Snowcloud snow depth data inferred from four deployed nodes is in Fig. 4. Processing includes some conservative noise removal, where sensor readings that are definitely spurious given known possible

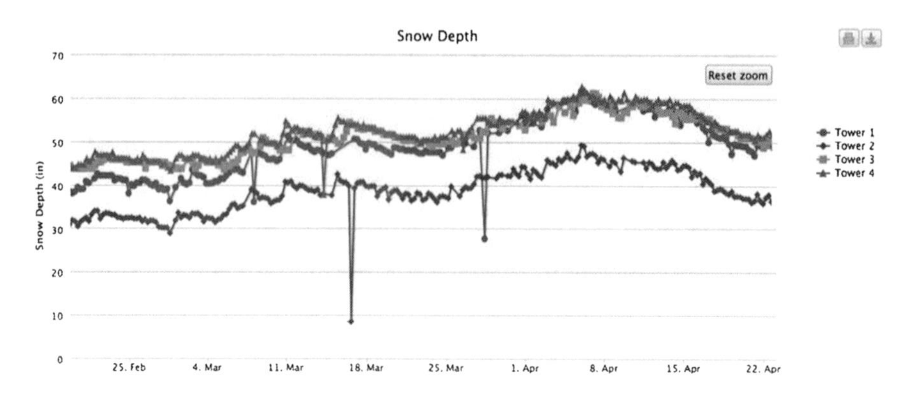

Fig. 4 Screenshot of snow depth data from Sulitjelma, Norway, 2013

value ranges are filtered out, otherwise smoothing is left to the end user. Processing also includes transformation of raw ADC voltage datapoints into physical units. These transformations depend on the sensors used and desired physical units. The air temperature and soil moisture sensors we've used come with factory-specified calibration curves for converting sensor voltages into physical units. Interpreting snow depth, PAR, and system voltages requires customized techniques since the relevant sensors are not "out of the box" for these applications. But for all sensors that we use, calibration curves are linear.

Snow depth Ultrasound sensors directly measure the time for a sonic pulse to travel from the sensor to a solid surface and back. Distance to the surface is easily inferred from this, though air temperature must also be known since the speed of sound varies with it. As the distance to ground surface G from any fixed sensor can be measured prior to snowfall, snow depth D is interpreted from an input temperature reading t in physical units and an ultrasound reading s in raw voltage as follows, where SOS is the speed of sound as a function of temperature, and C is the ultrasound calibration curve that converts raw voltage to time of flight: $D = G - ((C(s)/2) * SOS(t))$. The calibration curve C is not factory supplied, and ultrasound performance tends to vary, so each Snowcloud SD sensor array is calibrated individually to obtain a tower-specific C. This is done in lab conditions by recording sensor readings at defined distances and known temperatures, and performing a simple linear regression on the results.

PAR, V sensors Both PAR and system voltage readings are directly interpreted from sensor data. Although calibration curves must be obtained, we have found these curves to be quite consistent across sensor instances. For PAR sensors, we obtained a calibration curve for converting raw ADC counts to readings in micromoles/s^2, by plotting a set of ADC readings against PAR levels measured with a Decagon AccuPAR LP-80 ceptometer and performing simple linear regression. For system voltage calibration, we plotted voltage sensor ADC counts against input voltage levels, accurately set with a power supply, and performed a simple linear regression on the graph.

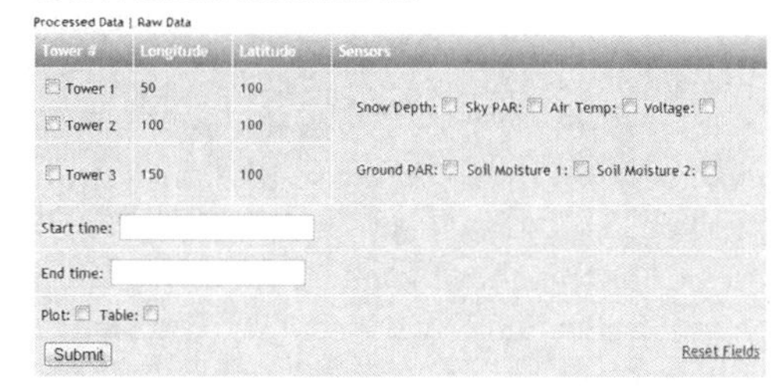

Fig. 5 Web-based user and administrative interfaces

Web Interfaces Both raw voltage and interpreted data is made available to users via online interfaces. A screenshot of the user interface for our Mammoth Lakes, CA deployment is shown in Fig. 5. The intent of the interface is to allow basic visualizations, and provide raw material for input into other tools, e.g., GIS. Thus, data can be presented in either graphical or tabular format. The graph in Fig. 4 is a screenshot of the web interface for a current deployment, and has interactive features online.

We have also developed web interfaces to improve administrative efficiency when setting up new sites, and maintaining existing ones. All software used in our system is largely consistent over various deployments, except processing scripts in particular are parameterized by the calibration curves used for deployed sensors. An administrative web interface allows calibration curves to be entered into the database and associated with specific sensor arrays for data processing.

5 Deployments and Field Experience

To date, we have deployed several Snowcloud systems to support scientists from several institutions. Deployment environments have included the Sierra Crest, the Eastern Sierra high desert, a New England forest, and arctic Norway. Thus deployment latitudes, altitudes, and climates have varied widely, as have research applications. This requires a highly adaptable, flexible, and robust data gathering system. Furthermore, these field experiences have motivated a number of refinements to our system hardware and software as discussed in the preceding text.

The deployments described here have succeeded insofar as usable datasets have been generated by each, and all but the Sagehen Creek dataset are available online at www.cs.uvm.edu/snowcloud. (Valid date ranges fall within deployment periods stated below.) Furthermore, analysis of this data reinforces the benefits of an automated, distributed system to capture highly variable snowpack properties [13]. As exemplified in Fig. 4, snowpack evolution typically exhibits clear spatiotemporal variability at different locations in deployments, so a distributed sensor system is well-suited for data gathering in this context. This evolution cannot be captured with the same temporal resolution using manual snow courses, or with the same spatial resolution using single-point measurement of a SNOTEL site.

Sagehen Creek Field Station, California, USA (Fall 2009–Spring 2010) Sagehen is situated just east of the Sierra Crest at an elevation of 2,000 m. The deployment period was December 2009 through June 2010. In addition to prototype testing, this deployment was used for collaborative research with University of Nevada, Reno (UNR). The results of this research demonstrated that the combination of telemetry obtained from a Snowcloud deployment, with models obtained using statistical techniques including linear regression and kriging, allows more accurate prediction of areal SWE averages than standard techniques [13]. This full-season field campaign served to validate basic functionality and robustness of the Snowcloud platform in its intended environment, and to demonstrate that our low-cost ultrasound-based approach to SD measurement specifically is effective.

The deployed network consisted of six sensor nodes, each supporting an aerial node with temperature and ultrasound sensors. The deployment covered a 1 hectare location with variety of terrain and canopy features. As a field research station, we were able to report data as it was collected, via the aforementioned collection tree protocol (CTP), to a base station mote connected to a laptop in a laboratory building proximal to the deployment site. As this laptop was connected to AC power and the Internet, data was available in near-real-time and data collection and reporting never failed. As we discuss in the subsequent deployments, such convenience in reporting is not the norm in practice.

Mammoth Lakes, California, USA (Winter 2012-date) An active Snowcloud network is currently deployed at an Easter Sierra Mountain site (elevation 2,300 m) near Mammoth Lakes. The data gathered by this network supports research directed by researchers from University of California, Santa Cruz (UCSC). The purpose of this research is to study the effects of climate change on alpine snow hydrology and high-desert flora, specifically the effect of increased rain-on-snow events on shrub communities. This deployment consists of three towers deployed over a 300 m transect, each with a fully-instrumented aerial node (SD, PAR and temperature) and a ground node (PAR, soil moisture at 10 cm and 1m). Leaf area index is derived from the difference between the PAR sensors in the aerial and ground arrays. Furthermore, the voltage sensor (discussed in technical detail in Sect. 2) provides useful *system* telemetry, i.e., an indication of battery levels over time. Both Harvester and Gateway device prototypes have been utilized for data collection in this deployment.

Hubbard Brook Experimental Forest, New Hampshire, USA (Fall 2012-date) and Sulitjelma, Norway (Winter 2013-date) During the past year we have deployed the Snowcloud system in two disparate but low altitude settings. The first site is on the forested slopes of the Hubbard Brook Experiment Forest in New Hampshire, USA (elevation 300 m). This area has been a site of a long term study to better understand snow and its impact on streams and watersheds. This particular deployment supports researchers from the University of New Hampshire (UNH) who are studying the effects of forest canopies on snow accumulation and melt. For this purpose we have installed three towers with aerial nodes to provide continual sampling at sites where manual snow courses are conducted nominally on a biweekly basis. Our second recent deployment is outside the town of Sulitjelma, Norway in collaboration with researchers at Stockholm University (SU). This site (elevation 150 m) is above the arctic circle which impacted greatly our ability to rely on solar for months during the winter, and gave our system its most extreme test to date. We have four towers with aerial nodes at this site and a two-month sample of the collected snow depth data can be seen in Fig. 4. The variability seen between towers deployed in near proximity (approximately 50 m apart) will help researchers develop more informative models for areal SWE for the purposes of validating airborne data. The Harvester device has been successfully used by our collaborators to retrieve data from both of these deployments.

6 Conclusion

In this chapter, we have described the Snowcloud system for snow hydrology research applications, that implements a complete data collection pipeline from *in situ* sampling to online presentation. The main novelty of the system is its application space, and its design support for strategic short- and medium-term studies and adaptability to a variety of missions. The system has been successfully deployed in harsh Winter conditions in a number of settings, demonstrating the robustness of its design and the effectiveness of distributed WSN technology for monitoring snowpack evolution.

As future work, we intend to expand applications of our system, and refine and deploy our Gateway technology during the upcoming 2014 snow season. We also intend to investigate network control algorithms to reduce system power consumption. These algorithms will leverage global knowledge and higher computing power on the Gateway, and will build on so-called backcasting techniques [19] for network control and new programming languages technology for control orchestration in WSNs [4]. Finally, we are working to augment Snowcloud with additional sensing capabilities including in situ temperature profiling and microwave attenuation to better characterize snowpack dynamics during melt onset.

Acknowledgments The authors express their thanks to our scientific collaborators, including David Moeser (Swiss Federal Institute for Snow and Avalanche Research—SLF), and Drs. Mark Walker (UNR), Michael Loik (UCSC), Jennifer Jacobs (UNH), and Ian Brown (SU). We would also like to thank Dan Dawson (Sierra Nevada Aquatic Research Lab—SNARL) and Jeff Brown (Sagehen Creek Field Station) for their invaluable support of our field work.

References

1. Alippi, C., Anastasi, G., Galperti, C., Mancini, F., Roveri, M.: Adaptive sampling for energy conservation in wireless sensor networks for snow monitoring applications. In: IEEE Conference on Mobile Ad Hoc and Sensor Systems (2007)
2. Campbell, J., Ollinger, S., Flerchinger, G., Wicklein, H., Hayhoe, K., Bailey, A.: Past and projected future changes in snowpack and soil frost at the Hubbard Brook Experimental Forest, New Hampshire, USA. Hydrol. Process. **24**, 2465–2480 (2010)

3. Campbell, D. et al.: Attribution of declining Western U.S. snowpack to human effects. J. Climate **105**, 11826 (2008)
4. Chapin, P., Skalka, C., Smith, S., Watson, M.: Scalaness/nesT: type specialized staged programming for sensor networks. In: ACM Generic Programming: Concepts and Experiences (GPCE) (2013)
5. Franz, K., Hatmann, H., Sorooshian, S., Bales, R.: Verification of national weather service ensemble stream-flow predictions for water supply forecasting in the Colorado River basin. J. Hydrometeorol. **4**, 1105–1118 (2003)
6. Guizzo, E.: Into deep ice. IEEE Spect. **42**(12), 28–35 (2005)
7. Hasler, A., Talzi, I., Tschudin, C., Gruber, S.: Wireless sensor networks in permafrost research—concept, requirements, implementation and challenges. In: Proceedings of 9th Int'l Conference on Permafrost (2008)
8. Henderson, T., Grant, E., Luthy, K., Cintron, J.: Snow monitoring with sensor networks. In: IEEE International Conference on Local Computer Networks (2004)
9. Kerkez, B., Glaser, S., Bales, R., Meadows, M.: Design and performance of a wireless sensor network for catchment-scale snow and soil moisture measurements. Water Resour. Res. **48**(9), w09515 (2012)
10. Lampkin, D.: Resolving barometric pressure waves in seasonal snowpack with a prototype-embedded wireless sensor network. Hydrol. Process. **24**, 2014–2021 (2010)
11. Liston, G.: Interrelationships among snow distribution, snowmelt, and snow cover depletion: implications for atmospheric, hydrologic, and ecologic modeling. J. Appl. Meteorol. Climatol. **38**, 1474–1487 (1999)
12. MEMSIC.: TelosB data sheet. Technical report, 6020–0094-03 Rev A (2013)
13. Moeser, C., Walker, M., Skalka, C., Frolik, J.: Application of a wireless sensor network for distributed snow water equivalence estimation. In: Western Snow Conference (2011)
14. NOAA.: Strategic science plan. Technical report, Office of Hydrologic Development Hydrology Laboratory (2007)
15. Tedesco, M., Narvekar, P.: Assessment of the NASA AMSR-E SWE product. IEEE J. Sel. Top. Appl. Earth Obs. Remote Sens. **3**(1), 141–159 (2010)
16. UNESCO.: Third UN world water development report: water in a changing world. Technical report, UN (2007)
17. Vilajosana, I., Llosa, J., Schaefer, M., Surinach, E., Marques, J.M., Vilajosana, X.: Wireless sensors as a tool to explore avalanche internal dynamics: experiments at the Weissfluhjoch snow chute. In: International Snow Science, Workshop (2009)
18. Welles, E., Sorooshian, S., Carter, G., Olsen, B.: Hydrologic verification: a call for action and collaboration. Bull. Amer. Meteor. Soc. **88**(4), 503–511 (2007)
19. Willett, R., Martin, A., Nowak, R.: Backcasting: adaptive sampling for sensor networks. In: IPSN, pp. 124–133 (2004)
20. Yan, F., Ramage, J., McKenney, R.: Modeling high-latitude spring freshet from AMSR-E passive microwave observations. Water Resour. Res. **45**, w11408 (2009)

The Big Night Out: Experiences from Tracking Flying Foxes with Delay-Tolerant Wireless Networking

Philipp Sommer, Branislav Kusy, Adam McKeown and Raja Jurdak

Abstract Long-term tracking of small-size animals with wireless sensor networks remains a challenge as only limited energy harvesting and storage is possible due to stringent size and weight constraints for animal collars. We present first experiences towards a perpetual monitoring system for free-living flying foxes. The high mobility of flying foxes requires a delay tolerant wireless network for data gathering: GPS positions and sensor data have to be stored locally until a wireless gateway deployed in bat congregation areas, so called roosting camps, comes within radio range. In this chapter, we present the system architecture and discuss our design decisions towards sustainable and reliable monitoring of flying foxes with a limited energy budget for sensing, storage and communication. Using empirical data from three free-living flying foxes, we characterize the overall system performance in terms of energy consumption and latency.

1 Introduction

The ongoing technological innovation, driven by the explosion of interest in mobile phone technology, has led to ever smaller, less energy demanding, and more accurate sensors and computation devices available on the market. Modern smart phones can

P. Sommer (✉) · B. Kusy · R. Jurdak
Autonomous Systems Lab, CSIRO Computational Informatics,
Brisbane, QLD, Australia
e-mail: philipp.sommer@csiro.au

B. Kusy
e-mail: brano.kusy@csiro.au

A. McKeown
CSIRO Ecosystem Sciences, Cairns, QLD, Australia
e-mail: adam.mckeown@csiro.au

R. Jurdak
e-mail: raja.jurdak@csiro.au

K. Langendoen et al. (eds.), *Real-World Wireless Sensor Networks*,
Lecture Notes in Electrical Engineering 281, DOI: 10.1007/978-3-319-03071-5_2,

localize users with an accuracy of a few meters, thus enabling novel applications that would not have been possible a few years ago. Decreasing form factor, weight, cost and energy consumption have made GPS receivers a versatile research tool enabling novel applications across several domains.

Wireless sensor networks (WSNs) are well suited for wildlife habitat monitoring applications. Sensor nodes, also called *motes*, combine sensing, processing, storage, communication, and energy harvesting capabilities in a small and light-weight package powered by batteries. Researchers typically place motes at specific locations to deliver non-intrusive long-term observation of natural habitats. However, many research questions require tracking the movements of individual animals within a population at high spatial and temporal resolution [3, 5, 9]. Recently, GPS-enabled collars weighting just above 100 g have been used for long-term tracking of Whooping Crane migration in North America [1]. As maximum weight of a collar is usually limited to 5 % of the body weight of the animal, tracking small-size animals remains a challenge.

Application context Flying foxes are mammals that belong to the family of fruit bats (*Pteropididae*). They are common in the tropic and subtropic areas of Asia, Australia and Pacific Islands. They congregate in large numbers of up to several thousands individuals to roost during daytime in so called *camps*. Flying foxes are nocturnally active and leave the camps for foraging from fruit trees. During the nightly foraging, they are able to cover distances up to 100 km over more than 10 h. Monitoring mobility and behavior of individual flying foxes is motivated by the need to better understand this threatened species. Furthermore, flying foxes are attributed to carry diseases, such as the Hendra or Lyssa viruses, which can be transmitted to other animals or humans.

Challenges Developing a hardware and software system to track flying foxes is challenging due to several constraints. First of all, animal ethics requires that the collar accounts for less than 5 % of the animal's body weight. Depending on the type of animal and differences in body weight between male and female individuals, this results in a total weight limit of 30 to 50 g for all electronic components, small solar panels and batteries. Consequently, we are severely limited in terms of available energy resources required to perform sensing, processing, storage and communication tasks. Second, our goal is to track free-living animals, so we will have no physical contact with the collar after the deployment time. Finally, flying foxes leave the camp for a period of more than 10 h every night and can be away for multiple days. Collars need to collect data in a delay-tolerant way and operate for long periods of time during which they cannot communicate with the base station.

Contributions In this work, we present the system architecture and discuss our design decisions to achieve perpetual low-power operation of a tracking device (Sect. 2). First, we have developed *BatMac*, a simple scheduling algorithm for low-power wireless communication tailored to our particular application domain. Specifically, a large number of animals congregate within the relatively small area of a camp during daytime, which allows for offloading previously gathered sensor data to a base station. We use a slotted schedule based on GPS time to desynchronize transmissions

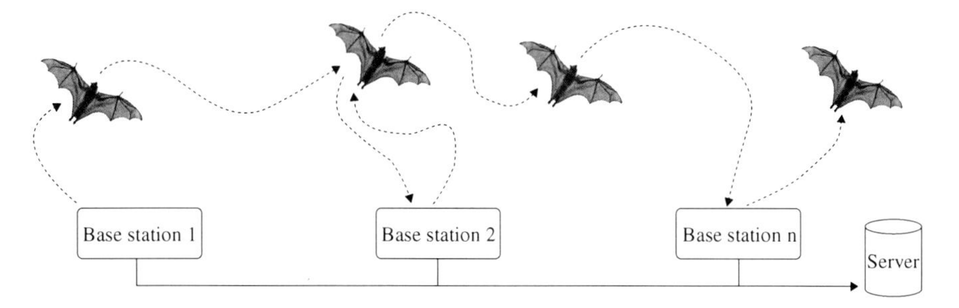

Fig. 1 The network architecture of the Bat Monitoring Project

of individual collars and use a simple radio duty-cycling approach that assumes an always-on base station (Sect. 3). Second, profiling and debugging of wireless sensor nodes on free-living animals paired with the intermittent radio connectivity is a challenging task. We have implemented a scheduler that guards execution of individual sensing tasks and use several mechanisms to improve software and hardware reliability. We also implemented an over-the-air reconfiguration protocol that helps to mitigate the lack of physical access to the device after the deployment time (Sect. 4). Finally, we use empirical data from three flying foxes to demonstrate performance of individual system components in real-world deployments.

2 System Architecture

In this section, we give a brief overview of the system architecture, shown in Fig. 1. Our wireless sensor network consists of three layers: (1) the mobile sensing nodes integrated into an animal collar deployed on flying foxes, (2) the base station layer which consists of several spatially distributed units, and (3) the central database server.

2.1 Mobile Sensing Layer

The purpose of the mobile sensing device is to gather sensor data from an individual collared animal using a variety of sensors (e.g., GPS, accelerometer, pressure sensor). The mobile sensor device is housed inside a collar, which can be attached around the animal's neck by experts trained in handling flying foxes.

In order to meet the stringent constraints in terms of weight, size and power consumption, we decided to build our own printed circuit board (PCB). A detailed description of this board is available in [8]. The software on the mobile sensor node is running a modified version of the Contiki operating system that adds custom extensions for logging and remote procedure calls (RPC) (see Sects. 3 and 4).

2.2 Base Station Layer

Bat roosting camps provide an ideal opportunity for the placement of static infrastructure, so called base stations, as thousands of animals congregate within a relatively small area during daytime. The base station is responsible for downloading sensor data from nearby mobile nodes by using short-range wireless connectivity. We use a gateway node with a TI CC1101 radio connected to an embedded Linux system for control and monitoring of the download operations. In addition, a 2G/3G wireless modem connects the base station to our central server for data uploads. We employ solar panels and batteries to allow autonomous operation in bat camps. Solar energy harvesting is usually a reliable source of power in tropical or subtropical locations with plenty of sunshine, but consecutive days with cloud cover or dense vegetation can limit the amount of solar energy harvested. Consequently, we might only be able to operate the base station during a limited time and have to batch downloading data from animals and uploading to the database.

2.3 Backend Storage and Control Layer

Sensor readings from different mobile nodes are downloaded by spatially separated base stations and transferred to a central database for permanent storage and offline analysis. The database is further responsible to keep a synchronized view of which pages have been already downloaded from nodes. This information is required to avoid duplicate downloads of the same page when the animal is roaming between different camps. Network health data such as battery voltage and number of packets received from different base stations are periodically reported to the database to assist in continuous monitoring and network management.

3 Delay Tolerant Networking for Animal Tracking

Flying foxes are known to cover large distances during nightly foraging and seasonal migrations between different camps. Satellite-based communication systems allow data upload at global scale but pose a significant burden in terms of their cost, size and power consumption. The large spatial coverage of cellular communication networks (e.g., 2G and 3G systems) offers a flexible and cost effective alternative to satellite based systems. However, size and power hinders the integration into collars for small mammals and birds with more stringent constraints. Therefore, we have opted for a low-power, short-range wireless transceiver (TI CC1101) that allows energy efficient operation within unlicensed bands of the frequency spectrum. The disadvantage of our approach is the need to maintain the infrastructure of base stations in flying-fox congregation areas.

Protocols for data collection During the last decade, several data collection protocols for wireless sensor networks have evolved for different application scenarios. Data collection protocols such as CTP [7] and Dozer [2], maintain a routing tree along which packets are forwarded towards the sink node. While both protocols are able to duty-cycle the radio transceivers to save energy, maintaining the network state requires periodic radio beacons. Recently, several communication protocols have been proposed that not need to keep topology-dependent state, such as the Low-Power Wireless Bus [6], which are resilient to high node mobility, but require the nodes to maintain accurate synchronization.

3.1 The BatMac Protocol

Given the uncontrolled mobility patterns of free-living animals and limited energy resources for wireless communication, we decided to implement a novel protocol called *BatMac*. BatMac is a time-synchronized medium access protocol, which is tailored to the intermittent connectivity between mobile nodes and the base station. BatMac is a sender-initiated single-hop protocol implemented on top of Contiki's RIME network stack. It is based on the observation that, unlike homogeneous sensor networks, the distribution of power budgets is highly asymmetric in our network. Mobile nodes can aggressively duty-cycle their radios while the base station operates its radio continuously. Therefore, we do not to use multi-hop communication for packet forwarding, which allows mobile receivers to put their radio in sleep mode for long periods of time.

Slotted communication We use a combination of time-based communication slots and a request-response protocol to avoid interference when multiple mobile nodes are present within communication range of a single base station. Medium access is scheduled using a concept of communication rounds that each consist of several sub-slots. Nodes can access the radio channel exactly once during each communication round, in a sub-slot that is determined by their node ID. For example, for a system with 10 nodes, we might define a communication round to take 5 min and consist of 10 sub-slots. Each of the 10 nodes will then transmit once every 5 min for up to 30 s.

Timings of rounds and sub-slots are determined based on UTC time tracked by the real-time clock of mobile nodes. The real-time clock is periodically re-synchronized on every GPS lock. The mapping between nodes and their corresponding sub-slot is based on their unique identifier. Selection of the round length and the number of sub-slots is a tradeoff between data transmission latency and the maximum number of nodes that we support. As our deployment needs to scale up to 1000 nodes or more, we can assign multiple nodes to the same slot through a modulo operation on their node IDs. We synchronize the real-time clock to within a second to the GPS time, which has the effect of randomizing transmission of beacons within the same sub-slot.

Announcement beacon Each mobile node periodically broadcasts a radio packet containing an announcement beacon at the beginning of its designated communication slot. The announcement contains the node's identifier, application version, and the current flash page number. After sending the radio packet, the node keeps its radio on until a predefined timeout (e.g., 1 s) expires. If the timeout expires, the radio is switched off until the next announcement.

Node selection The base station is continuously listening for incoming announcement beacons from mobile nodes. Upon reception of an announcement, the base station determines if further communication with the node is required, e.g., if new data needs to be downloaded from the flash or the node's configuration should be updated. If further communication is needed, the base station keeps the node's radio awake, by sending a `radio_on(timeout)` command, which will set a new timeout to switch off the radio at the node.

3.2 Data Storage

We are interested in collecting sensor readings from different sensors while the collar is on the animals for several weeks, months or years. Mobility patterns of free-living animals make it very difficult to predict when the animal will be nearby a camp where a base station is deployed. Thus, our software is required to provide persistent storage of sensor data for several hours, days or even weeks. We implemented a first-in first-out data store using the external flash as a circular buffer. Our AT25DF641 flash chip is divided in 32768 pages of 256 bytes each.

The variety of sensors on the mobile device requires a data storage system that is able to handle readings of different payload sizes and at different data rates. For example, a single GPS reading includes values for the timestamp, latitude, longitude, height, speed and an estimation of position accuracy, which results in a total of 15 bytes every second. On the other hand, the combined 3-axis accelerometer and magnetometer sensor generates 12 bytes per reading and can operate at sampling rates up to 100 Hz.

Tagged data format We adapt the Tagged Data Format (TDF) from [4] to pack sensor readings into a byte stream. TDF adds metadata such as the sensor type and timestamp in front of each reading (see Fig. 2). The ID of the sensor type and flags indicating the type of timestamp (relative or absolute) are encoded into a 2-byte header field. Sensor readings associated with an absolute timestamp need 6 bytes to encode the seconds (4 bytes) and millisecond fraction (2 bytes) of the timestamp. If the timestamp of the current reading can be encoded using an offset to the previous timestamp, it is only necessary to store a 2-byte offset. Each sensor type has a specific length for sensor readings, which is fixed and needs to be known to both the encoder and decoder of a TDF stream. To enable decoding of each flash page individually, the first sensor reading always uses an absolute timestamp.

Sensor Type + Flags 2 bytes	Timestamp (global) 6 bytes	Sensor Data 1..n bytes	Sensor Type + Flags 2 bytes	Timestamp (offset) 2 bytes	Sensor Data 1..n bytes	Sensor Type + Flags 2 bytes	Timestamp (offset) 2 bytes	Sensor Data 1..n bytes	

Fig. 2 Storage of sensor readings in flash using the tagged data format (TDF)

Evaluation TDF is a flexible format that provides a compact representation of heterogeneous sensor readings in flash storage. However, the TDF encoder has to jump to the next page if not enough bytes are available in the remainder of the flash page. This fragmentation leads to empty bytes at the end of a page. We characterize the overhead of encoding sensor readings using TDF on sensor data from a collar attached to a free-living flying fox. We downloaded 1147 pages from the flash storage of the node and analyzed 17730 sensor readings encoded in the TDF stream. Our results indicate that the actual sensor data accounts for 65 % of the flash page size of 256 bytes, while headers account for 12 % (sensor type) and 19 % (timestamp). Finally, the overhead due to empty bytes at the end of a flash page accounts for only 4 % of the flash size, which is acceptable given the flexibility that TDF offers for storing heterogeneous sensor data into a continuous flash buffer.

3.3 Data Retrieval

Data downloads are initiated by the base station as a response to an announcement beacon, based on the node's current flash page number contained in the beacon. The download handler runs as a Python script on the embedded Linux machine. We implemented a greedy approach to scheduling downloads from nodes within radio range of a base station. If the current page number in the beacon is higher than the last downloaded page, the base station requests missing pages from the node in sequential order.

Since a full page of 256 bytes would not fit into a single radio packet, we use RPC calls to request chunks of bytes within a specific page from the base. Each RPC command is retransmitted up to 5 times if not acknowledged by the node. If a complete page has been transferred, the base reassembles it and decodes its content using the TDF parser. If the page contains no errors, the base will upload the sensor data to the central database server and mark the corresponding page as complete. If no Internet connection is available at the base station, data will be buffered locally at the base until it can be uploaded to the database.

Data consistency The request/response approach requires the base station to know the latest information from each node that is stored in the database. Since our system architecture includes multiple spatially distributed base nodes, we use a centralized approach to coordinate data downloads across the base stations. Base stations keep track of the most recently downloaded flash page for every mobile node, which allows mobile nodes not to keep track of the download process. Mobile nodes need to simply announce their most recent flash page number and then respond to a base

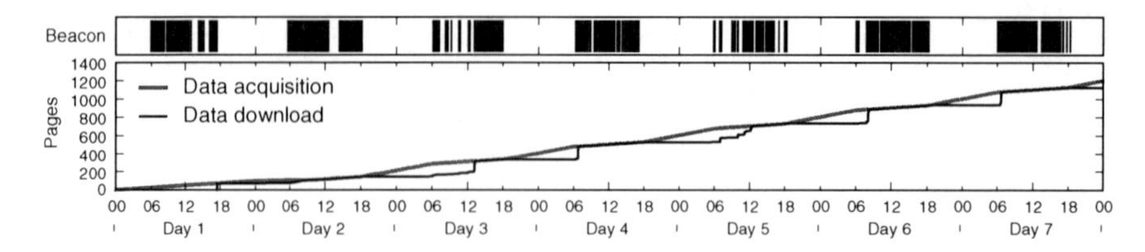

Fig. 3 Timeline of received radio announcement packets at the base station from a single bat (*top*) and the number of generated versus downloaded flash pages (*bottom*)

station's RPC to transmit a specific flash page. Each page will only be downloaded exactly once as long as the state information is synchronized across all base stations. However, this needs to be done only every couple of hours as animals are unlikely to move between the spatially separated base nodes within that time window.

Evaluation We present experimental results using data downloaded from a single wild bat during a 7-day period. Figure 3 shows received announcement packets and number of pages written to flash. In general, we are able to receive announcement beacons from the mobile node just before 6 am in the morning until just after 6 pm in the evening. In the morning, the earliest received beacon was at 5.34 am on the second day while no beacons were received until 6.14 am on Day 4. The last beacon from the animal was received between 5.09 (Day 1) and 6:19 pm (Day 5, 6 and 7). We calculate the packet reception rate (PRR) for announcement beacons as the fraction between received beacons and the number of expected beacons between the first and last received beacon for every day. The range of the observed PRR is between 0.35 (Day 5) and 0.78 (Day 2).

By analyzing the timestamps embedded in the downloaded pages, we are able to track the flash storage consumption over time. Flash pages are written at a lower rate during daytime, as we are mainly logging node health information and a few GPS positions. The data rate increases between 6 pm and 6 am when high-frequency GPS sampling is activated. The latency between data acquisition and download is low during the day since the base station is able to download new pages continuously. Clearly, the latency is higher for data gathered during the night as we have up to 12 h without contact to the base station. Depending on the quality of the radio link in the camp, it might take several hours until the nightly backlog is downloaded (e.g., Day 5).

4 Configuration and Debugging

Development of hardware and software for animal tracking is challenging due to the mobility of the animals. While it is relatively easy to follow the path of collared livestock, catching and collaring of free-living animals such as flying foxes is labor intensive and notoriously difficult. Large nets mounted between high poles are required to catch animals while they are flying back in or out of the camp during the night. Catching a collared animal a second time is almost impossible given the large number of individuals populating a camp. Therefore, our development approach assumes that it is not possible to gain physical access to the mobile node ever again after the initial deployment.

4.1 Remote Task Configuration

Several mechanisms for over-the-air code distribution in wireless sensor networks have been proposed in the literature. However, supporting wireless reprogramming increases the complexity of the code running on the node and requires dedicated storage for new and fallback images. Furthermore, any failure during the reprogramming process might leave the node in a defective state. Therefore, we decided not to implement over-the-air reprogramming for our mobile nodes. Instead, we integrated methods to support wireless reconfiguration for a set of well-tested tasks within our application. Each task is associated with a Contiki process that implements a specific sensing task (e.g., getting several GPS fixes, or measuring the battery voltage). Tasks can be limited to a specific time interval (start, stop), periodicity within that time interval, and minimum battery voltage. Tasks can also have additional arguments which are specific to a sensor (e.g., number of samples). The task scheduler is executed once every second to start or stop tasks according to the current configuration.

Reconfiguration Task configurations use a dedicated part of the flash for persistent storage. We provide remote procedure calls (RPC) sent over the radio to view, update and delete a task on the sensor node.

4.2 Remote Debugging

We implemented two methods for debugging mobile nodes deployed on animals: Node inspection by remote procedure calls sent over the radio, and logging of debug output to the flash storage as part of TDF data. In addition, we use a combination of hardware and software based mechanisms to recover from error conditions. In the remainder of this section, we describe the debugging capabilities integrated with our application and highlight how their usefulness for debugging in practice.

Node inspection We implemented several helper methods for debugging and inspection of the current node state. Thereby, we do not want to halt code execution on the node, but rather acquire a snapshot of the node's status, e.g., the current value of a variable in RAM. Furthermore, custom RPC methods allow us to reboot the node, sample the battery voltage, read data from the external flash, and read/modify/delete tasks.

Debug instrumentation While RPC methods are useful to inspect the node status when radio connectivity is available, little information is available during periods with no radio connectivity. Therefore, we implemented two additional sensing tasks useful to aid the process of debugging. The *POWER* task will periodically sample the battery voltage, solar panel voltage, and charge current and write the ADC readings to the flash storage using the TDF logging abstraction. In addition, the *DEBUG* task periodically logs the reason for the last microcontroller reset and the current uptime in seconds, allowing us to track node reboots and determine their cause retrospectively.

Watchdog and Grenade timers We combine a hardware watchdog and a software timer to recover from potential problems that could cause the software running on the microcontroller to get stuck. The hardware watchdog of the MSP430 will trigger a reboot if the watchdog timer is not being reset within 2 s. Therefore, we instrument Contiki's main scheduler loop to reset the watchdog periodically. This mechanism will protect us from possible software errors within the implementation of our Contiki processes such as infinite loops. A so called *grenade timer* is used as an additional layer of protection against hardware and software problems. Thereby, we force a graceful node reboot when the uptime counter reaches a predefined value (e.g., once a day). Grenade reboots can easily be distinguished from uncontrolled reboots by looking both at the reset reason and the value of the uptime counter logged to flash.

5 Experimental Data: A Day in the Life of a Flying Fox

We present preliminary experimental results from two mobile nodes attached to free-living flying foxes. Both animals left the camp during the night and returned at dawn. We configured two different tasks for GPS sampling to gather the position of the mobile node depending on the current time of day. The low-frequency GPS task will log the current location every 3 h during the day. During the night, the high-frequency GPS task logs locations every 10 min.

5.1 GPS Tracking

The interval between startup of the GPS receiver until the first GPS position is available is commonly denoted as the *Time to First Fix (TTFF)*. During a *coldstart*, the GPS receiver has to search for a signal from several GPS satellites to determine its current position. After the first valid position has been calculated, the receiver keeps tracking the existing satellites in order to periodically update its estimation of the current location. We use the sleep mode of the u-blox MAX6 module, which retains real-time clock and satellite orbit information while the main part of the receiver is put into a low-power mode. A GPS *hotstart* is possible if the receiver wakes up from sleep mode and has up-to-date satellite information to calculate the current position, otherwise, it has to go through a coldstart phase again. We collected TTFF values of 1103 successful GPS location requests from two nodes. Reported values for TTFF

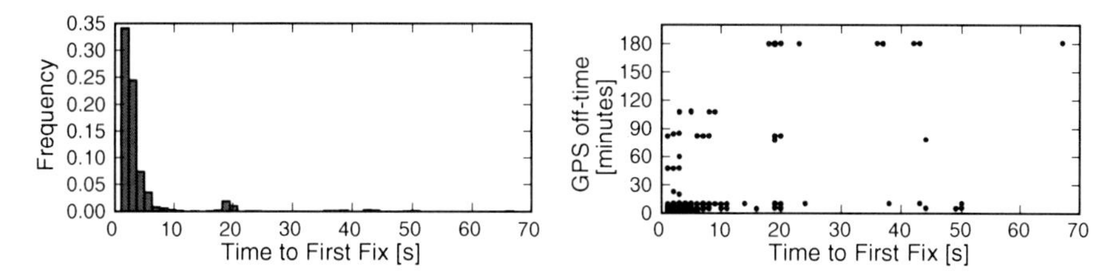

Fig. 4 Distribution of the time to first GPS fix (*left*) and the relation between the preceding GPS off interval and the time to first fix (*right*)

are between 1 and 67 s while it takes 4.17 s to get the first valid position on average (see Fig. 4). We further observe that it takes longer to get the first fix when the GPS was inactive for a longer period of time.

5.2 Power Management

The amount of remaining energy stored in the rechargeable battery depends on several factors, such as the amount of solar energy harvested and the current draw during recent days. We periodically measure the battery voltage, the solar panel voltage and the charge current into the battery using the on-board ADC on the sensor node. While we aim to derive an accurate estimation for the remaining energy based on measurements in future versions, the task scheduler currently only uses the battery voltage when deciding whether to execute a task. Power measurements and corresponding GPS task execution times are reported in Fig. 5. We measure similar values for the charge current for both nodes. We see considerable differences between the two mobile devices, although the animals are roosting in the same camp. Node A maintains a relatively stable battery voltage across the whole measurement period while the battery voltage of Node B decreases rapidly during the night time. As a result of the low voltage, Node B stops gathering GPS samples as its battery drops below the threshold voltage. We can also see a significant difference between rainy weather (Day 1) and sunny weather (Days 2 to 5). Furthermore, we continually increase the duty-cycle of the GPS task during the night. Starting from fixes every 10 min between 6 pm and 6 am (Day 1/2), we extended the window from 5.30 pm to 6 am to capture the departure of the animals from the camp (Day 3/4). Finally, we reconfigured the GPS task to gather continuous GPS fixes at a frequency of 1 Hz between 5.35 and 6 pm on the evening of Day 5. Figure 6 shows the positions of the animals on a map.

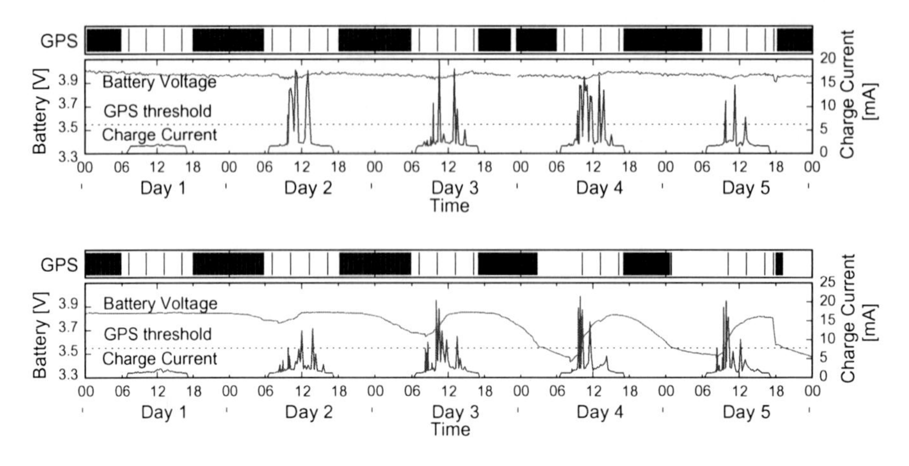

Fig. 5 Battery voltage, charge current and activation of the GPS task for Node A (*top*) and B (*bottom*)

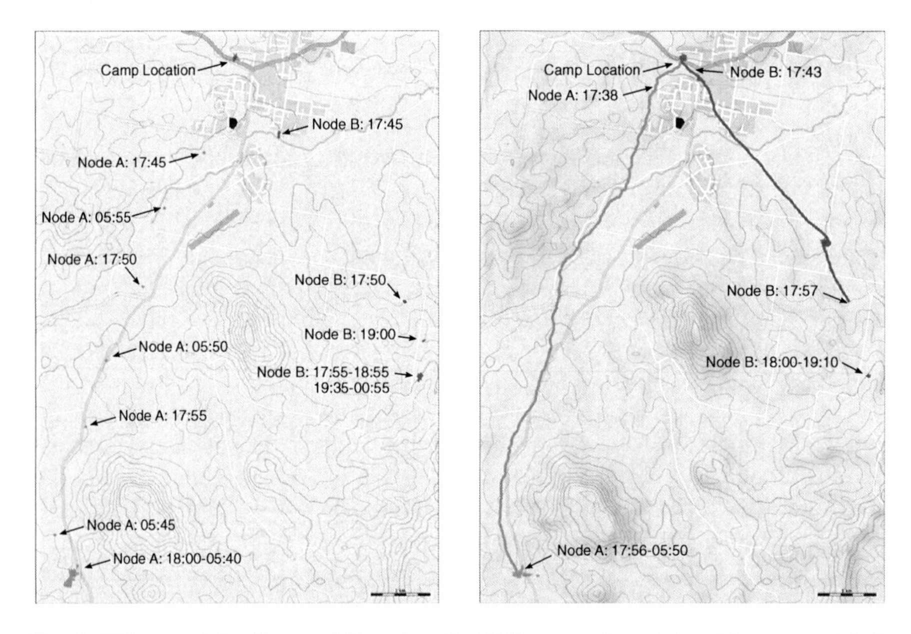

Fig. 6 GPS waypoints of two mobile nodes: The GPS was configured to get fixes every 10 min during night time on Day 4 (*left*). Continuous sampling allows to track the accurate flight path when animals leave the camp on Day 5 (*right*)

6 Lessons Learned

In this chapter, we have presented our first experience with a novel hardware and software architecture for monitoring flying foxes using mobile sensor nodes. Although this project is only at an early stage and deployment of several hundred animal collars is planned over the next months, we have gathered a large set of empirical data, which allows us to verify the correctness of system operations and provides invaluable information for fine-tuning our system in future deployments.

We believe that staggered releases of new software features to a small number of nodes can mitigate the impact of possible software problems instead of rolling out several hundred nodes simultaneously. However, this requires co-existence of different software versions, as collars with older software versions still remain active. Thus, we assign a software version to each node which is also included as part of the announcement beacon, enabling to identify the capabilities of different nodes. Furthermore, diagnostic tools built on top of remote procedure calls enable flexible inspection of nodes within range.

Logging state information to persistent storage is vital for debugging a system with intermittent network connectivity. The over-the-air reconfiguration of tasks provides control over the amount of debug output during different stages of the deployment, thus providing code instrumentation only when needed.

Acknowledgments We would like to thank Ben Mackey, Philip Valencia, Chris Crossman, Luke Hovington, Les Overs and Stephen Brosnan for their contributions to this project.

References

1. Anthony, D., Bennett, W., Vuran, M., Dwyer, M., Elbaum, S., Lacy, A., Engels, M., Wehtje, W.: Sensing through the continent. In: ACM/IEEE IPSN (2012)
2. Burri, N., von Rickenbach, P., Wattenhofer, R.: Dozer: Ultra-low power data gathering in sensor networks. In: ACM/IEEE IPSN (2007)
3. Butler, Z.: From robots to animals: virtual fences for controlling cattle. J. Robot. Res. **25**, 5–6 (2006)
4. Corke, P., Wark, T., Jurdak, R., Moore, D., Valencia, P.: Environmental wireless sensor networks. Proc. IEEE, **98**(11), (2010)
5. Dyo, V., et al.: WILDSENSING: design and deployment of a sustainable sensor network for wildlife monitoring. ACM Trans. Sens. Netw.**8**(4), (2012)
6. Ferrari, F., Zimmerling, M., Mottola, L., Thiele, L.: Low-power wireless bus. In: ACM SenSys, (2012)
7. Gnawali, O., Fonseca, R., Jamieson, K., Moss, D., Levis, P.: Collection Tree Protocol. In: ACM SenSys, (2009)
8. Jurdak, R., Sommer, P., Kusy, B., Kottege, N., Crossman, C., Mckeown, A., Westcott, D.: Camazotz: multimodal activity-based GPS sampling. In: ACM/IEEE IPSN, (2013)
9. Zhang, P., Sadler, C.M., Lyon, S.A., Martonosi, M.: Hardware design experiences in ZebraNet. In: ACM SenSys (2004)

On Rendezvous in Mobile Sensing Networks

Olga Saukh, David Hasenfratz, Christoph Walser and Lothar Thiele

Abstract A rendezvous is a temporal and spatial vicinity of two sensors. In this chapter, we investigate rendezvous in the context of mobile sensing systems. We use an air quality dataset obtained with the OpenSense monitoring network to explore rendezvous properties for carbon monoxide, ozone, temperature, and humidity processes. Temporal and spatial locality of a physical process impacts the number of rendezvous between sensors, their duration, and their frequency. We introduce a rendezvous connection graph and explore the trade-off between locality of a process and the amount of time needed for the graph to be connected. Rendezvous graph connectivity has many potential use cases, such as sensor fault detection. We successfully apply the proposed concepts to track down faulty sensors and to improve sensor calibration in our deployment.

1 Introduction

In recent years, wireless sensor networks (WSNs) have become a mature technology and are successfully used in a number of long-term installations [1, 2]. Most deployments are composed of static sensors fixed at desired locations, which are carefully chosen according to the node's communication and sensing ranges. The data measured by such deployments is usually highly temporally resolved at every chosen

O. Saukh (✉) · D. Hasenfratz · C. Walser · L. Thiele
Computer Engineering and Networks Laboratory, ETH Zurich, Zurich, Switzerland
e-mail: saukh@tik.ee.ethz.ch

D. Hasenfratz
e-mail: hasenfratz@tik.ee.ethz.ch

C. Walser
e-mail: walser@tik.ee.ethz.ch

L. Thiele
e-mail: thiele@tik.ee.ethz.ch

K. Langendoen et al. (eds.), *Real-World Wireless Sensor Networks*,
Lecture Notes in Electrical Engineering 281, DOI: 10.1007/978-3-319-03071-5_3,

location. The main drawback of static installations is, however, their poor spatial scalability to monitor large areas. Involvement of mobile sensor nodes increases spatial area coverage at the price of temporal coverage. Additionally, it enables aperiodic *rendezvous* between sensors. We define a rendezvous as a temporal and spatial vicinity of two sensors. A rendezvous is not bounded to a specific location in time and space. In fact, two sensors moving around hand in hand build one rendezvous with infinitely many temporally and spatially related measurements.

Despite the advantages of mobile sensors, ensuring faultless operation [3, 4], automatic sensor calibration [5, 6], and data quality (e.g., by timely detection of outliers [7]) are challenging tasks in mobile networks due to the sporadic and unpredictable nature of rendezvous. Additionally, access to reliable reference measurements is absent or limited to a small number of locations. Thus, supervising the error-free operation of mobile sensor networks is only possible by either carefully designing and exploiting rendezvous between sensors, or applying model-based data validation schemes. The latter requires models describing the phenomenon of interest and sensors in use, which are typically not known a priori [8] and hard to obtain [9, 10].

In many urban areas, public transport vehicles, such as streetcars and buses, provide a suitable basis for deploying wireless sensor nodes [11]. The inherent mobility in such installations often solves spatial coverage problems and enforces high probability of rendezvous due to a usually good connectivity of public transport networks (PTNs). We monitor airborne pollutants and environmental parameters with the OpenSense nodes, which are installed on top of public streetcars traversing the city of Zurich, Switzerland.

In this chapter, we use our deployment to implement an in-depth analysis of rendezvous between mobile sensor nodes to ensure consistent network operation, and investigate prerequisites and limits of this approach. To the best of our knowledge, this is the first chapter researching this direction. The presented results are based on data obtained from three months of operation gathering more than 4 millions measurements.

Past research efforts on exploiting rendezvous in mobile networks focused on data communication [12–15] and network-wide time synchronization [16]. In both cases, a rendezvous depends on the node's *communication range* and is assessed in terms of successful communication between two nodes. This definition is *binary*, since communication either succeeds or fails. Many environmental monitoring applications would leverage a phenomenon-based definition of rendezvous for their use in sensor calibration and sensor failure detection routines [17, 18]. However, *given temporal and spatial distances between two sensors, what can be concluded about the similarity of their respective measurements? How close should two sensors come together so that one can expect their measurements to be similar?* In this chapter, we make use of our dataset to shed light on these questions.

Contributions and Roadmap Different phenomena exhibit different notions of spatial and temporal locality. Once two sensors are in close vicinity of each other, they are expected to deliver similar measurements. In this chapter, we propose a phenomenon-based definition of a rendezvous, which exploits this property using air pollution measurements from our mobile OpenSense nodes traversing a large urban area of 100 km^2.

The main contributions described in this chapter are:

- We introduce the notion of a rendezvous based on the spatial and temporal locality of a phenomenon. Using the OpenSense dataset, we can clearly distinguish global (temperature and humidity) from more local (carbon monoxide) phenomena.
- We construct a rendezvous connection graph, which enables comparing any pair of sensors in the network. We explore the trade-off between locality of a phenomenon and time required to achieve connectivity in the rendezvous connection graph. We find that in our deployment connectivity is achieved within a few days even for local phenomena, such as carbon monoxide dispersion in an urban environment.
- We demonstrate the applicability of rendezvous to detect sensor failures and to improve sensor calibration in the running OpenSense deployment.

We summarize related work in Sect. 2. In Sect. 3 we present the OpenSense deployment and describe the dataset used in Sect. 4 to explore the properties of rendezvous. The determined rendezvous parameters for a number of environmental phenomena are used in Sect. 5 to detect faulty sensors and update sensor calibration. We conclude in Sect. 6.

2 Related Work

Two directions in WSN research are related to the contributions of this chapter: exploiting node rendezvous in mobile networks and characterization of sensors' sensing ranges. Many applications of mobile networks leverage node rendezvous for data communication [12] and time synchronization [16]. A large body of work explores temporal connectivity in opportunistic networking, e.g. [19]. Several works advocate data muling to minimize node energy consumption by gathering data with a mobile sink [13–15]. The resulting communication range based rendezvous are defined by a successful communication link and are thus binary in contrast to sensing range based rendezvous.

Unit disc graph model [20] of a sensing range is a popular assumption in a great body of theoretical publications. Although this model is rather simplistic, it is widely used for theoretical analysis and algorithm design. In particular, many early coverage and k-coverage scheduling algorithms are based on this model [17, 21–24]. These works often either ignore the deviation of the model from reality or assume that the sensing radius can be obtained. A few works even propose to calibrate the sensing range to fit specific requirements [25, 26].

More recent research activities explore diversity of sensing ranges and build algorithms that are not limited to any specific model. In [27], the authors developed an energy efficient node scheduling algorithm for environmental monitoring applications, which exploits correlations between sensors to create the maximal number of uncorrelated subgroups. Energy savings are achieved by keeping active only one node from each subgroup at a time. The authors of [28] explore the impact of sensing diversity on sensor collaboration with the goal to achieve confident sensing coverage. The study of sensing diversity is based on a sensor network deployment for vehicle detection. The main idea of [29] is to develop precise models of a sensing range by relating the location of events to event detection results of individual sensors. Events are either generated intentionally or highly dynamic natural events are used to learn the sensing ranges. Similarities of measurements are computed using $\mathcal{P}$-norm, that, in contrast to our work, needs the sensors to be well-calibrated. Another closely related work is [3], that discusses simulation results of a neighbor-based node fault detection scheme by comparing sensor measurements within the node's neighborhood. Similar to our work, the authors propose to apply smooth filtering to correct some false readings due to transient faults.

In contrast to previous research, we investigate the problem of rendezvous design based on *unknown* sensor sensing range in a *mobile* scenario and in an *uncontrolled* setting. Our analysis and the obtained rendezvous parameters have not only spatial but also *temporal* characteristics, which is not explored in any of the above works. Our approach works for noisy uncalibrated sensors and can be used for detecting sensor faults and sensor calibration.

3 OpenSense System and Deployment

OpenSense is an air quality monitoring network composed of mobile sensor nodes, which leverage mobility of a local public transport network (PTN), and two governmental static stations providing high-quality measurements. The operation cycles of the PTN underlying the OpenSense network impact rendezvous times and locations. In this section, we present the OpenSense node's hardware and software as well as statistics on the OpenSense network operation. The evaluation is based on a three-month dataset acquired from March to May 2013. This dataset is used in the next section to examine the properties of rendezvous for several environmental phenomena.

The core of an OpenSense node, depicted in Fig. 1, is a Gumstix embedded computer with a 600 MHz processor running the Ångström embedded Linux operating system. A GPS receiver supplies the station with precise geospatial information. The weight of a node is approximately 4.5 kg and its power draw is around 40 W. Variations of the local air quality is measured with a set of gas sensors measuring ozone (O_3) [30] and carbon monoxide (CO) [31], along with temperature and relative humidity [32]. Five OpenSense nodes are equipped with a second temperature sensor located inside the box. All sensors face the direction of vehicle movement and are protected with covers from dust and rain. A fan on the back of the sensor box ensures a constant air flow through the box. O_3, temperature, and humidity are measured every 20 s while the CO and internal temperature sensors are sampled once every 60 s. All measurements are transmitted over the GSM network to the back-end server running GSN [33], a software middleware that facilitates data collection in sensor networks.

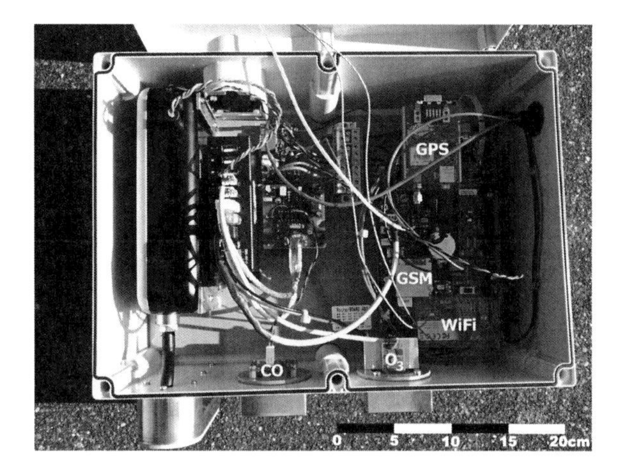

Fig. 1 The OpenSense node is equipped with several sensors, such as ozone (O_3) and carbon monoxide (CO), to monitor airborne pollutants and environmental parameters

To prevent data loss due to communication failures, the OpenSense node temporary stores measured data in a local database until its reception is acknowledged by the back-end server.

Two reference stations, part of the governmental local air quality monitoring network, are located 4 and 16 m apart from the PTN tracks. The stations are equipped with several high-quality analytical instruments capable of measuring main air pollutants with high precision. Data from these stations has a temporal resolution of 60 s.

Since February 2013, ten OpenSense nodes are operating on top of streetcars by traversing an urban region of 100 km^2. Figure 2b, c show operation times of each node. A sensor node operates approximately 20 h per day. During the night, typically from 1:00 AM to 5:00 AM, the streetcars are in their depots. Not all streetcars are used every day. On a regular day, on average, nine out of the ten streetcars are in operation. The streetcars are hosted by a set of depots in the city marked with circles in Fig. 2a. A depot serves a subset of lines, thus, any of its routes can be taken by a streetcar hosted by the depot. A streetcar changes routes from time to time or remains inactive in a depot for a few days. For example, node 1 remained in the depot for several weeks after an accident, as observable in Fig. 2b on the low number of days in operation. Although sensor mobility is limited to streetcar tracks, the OpenSense network covers many places with low and high traffic densities, parks, shopping and living areas. In total, a sensor node monitors air pollution along 3–6 different routes. On average, a streetcar decides daily on a new route. This behavior gives two advantages with respect to air quality monitoring on top of a streetcar network: (i) good spatial and temporal coverage of the city, as shown in Fig. 2a and (ii) high chances of rendezvous between sensor nodes or between sensor node and reference stations within a few days.

During the 3 months under analysis, the ten sensor nodes transmitted in total over 4 million data packets with O_3, CO, temperature, and humidity readings, as depicted in Fig. 2d. We use this data to investigate the temporal and spatial locality of the

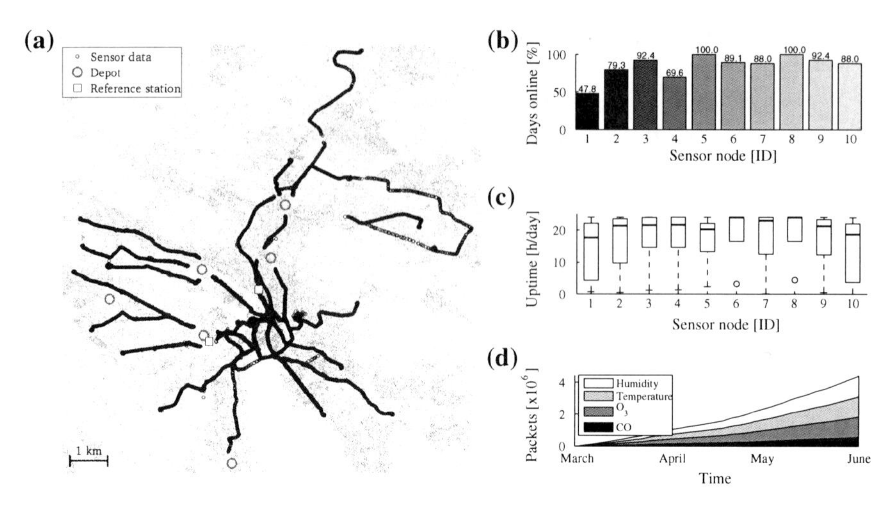

Fig. 2 OpenSense nodes cover a large urban area of 100 km^2 on *top* of ten streetcars. On average nine sensor nodes are in operation for 20 h per day. In three months more than 4 million measurements are performed. **a** Coverage of the urban area, **b** percentage of days in operation, **c** uptime per day, **d** packets over time

measured processes and to determine relevant rendezvous parameters. We ensure that the GPS position and the timestamp of each sensor reading are accurate. To this end, our system implements two data filters: (i) a *GPS-Filter* to eliminate sensor readings with inaccurate GPS locations and (ii) an *Indoor-Filter* to filter out data measured inside depots. About 57.1 % (2.46 million) of the measurements pass the two data filters.

The *GPS-Filter* leverages the horizontal dilution of precision (HDOP) value, which indicates the geometrical positioning of the GPS satellites. Smaller values imply a good positioning. In our implementation, we filter out sensor readings with HDOP greater than 2.5, which results in 1.1 % eliminated values. The 2.5 threshold corresponds to a localization precision of a few meters. Later we will show that the processes measured with the OpenSense network require spatial closeness of tens of meters and thus can tolerate position inaccuracies of several meters.

The *Indoor-Filter* eliminates further 41.8 % of measurements that were possibly performed in depots. Although the measurements are generally valid, they do not provide close estimates of the urban air pollution and lead to inconsistent (indoor/outdoor) value pairing at a rendezvous. *Indoor-Filter* removes all measurements taken by an OpenSense node within the radius of 300 m from the depot location (as can be observed in Fig. 2a by missing measurements in the close vicinity of depots). It also eliminates sensor measurements taken up to 30 min after leaving the depot to account for the time needed to adapt the sensor signal to outdoor conditions.

4 Rendezvous and Graph Connectivity

Let u and v be two, possibly mobile, sensors. Both sensors take time-ordered measurements $\{x_i\}, \{y_j\} \subset D, \forall i, j \in \mathbb{N} \setminus 0$ from a domain of sensor values D. Let Φ be a set of pairs of measurements taken by the sensors u and v defined by

$$\Phi = \{(x_i, y_j) \mid d(x_i, y_j) \leq \hat{d} \wedge t(x_i, y_j) \leq \hat{t}\},$$

where $d(x_i, y_j)$ and $t(x_i, y_j)$ define temporal and spatial distances between the measurements x_i and y_j, respectively. The parameters $\hat{d}$ and $\hat{t}$ set constraints on the closeness of two measurements. The sensors u and v make a *rendezvous* if the set of measurement pairs includes at least k values, $|\Phi| \geq k$, where k depends on the use case of the set Φ as we will show later. Pairs of measurements in Φ are referred to as *rendezvous pairs*. A rendezvous pair can be determined by comparing timestamps and GPS positions of two sensor readings taken by u and v. Rendezvous allow a number of useful applications, such as detecting sensor failures and performing sensor calibration, if correlation of rendezvous pairs is high. In this section, we experimentally investigate the dependency between rendezvous parameters $\hat{d}$ and $\hat{t}$ and correlation of rendezvous pairs for different environmental processes based on the sensor measurements acquired by the OpenSense deployment. A process is *local*, if it requires tightly set rendezvous parameters to achieve high correlation of rendezvous pairs. We investigate the trade-off between locality of a process of interest and amount of time required to get sufficient number of rendezvous pairs between sensors in the OpenSense network to draw conclusions on the status of the deployed sensors.

4.1 Process-Induced Rendezvous

We expect spatially and temporally close sensor readings to be similar [3, 27]. Given a set of rendezvous pairs for some rendezvous parameter values $\hat{d}$ and $\hat{t}$, we compute the Pearson correlation coefficient to quantify their similarity. We require $|\Phi| \geq 1{,}000$ rendezvous pairs between any two sensors and vary the values of spatial and temporal rendezvous parameters $\hat{d} \in [25, 500\,\text{m}]$ and $\hat{t} \in [1.5, 30\,\text{min}]$. Although not all sensors used in the evaluation are perfectly calibrated, the calibration curves of these sensors are nearly linear [30, 31]. In this case, insufficient calibration has no impact on the correlation between two sensors.

In the first row of Fig. 3, we present the average pairwise correlation of rendezvous pairs for the processes monitored by the OpenSense network: CO, O_3, temperature, and humidity. The obtained results suggest high variability of sensing patterns both in time and space depending on the process of interest. We observe an irregular correlation decrease pattern whereas our expectation is a monotonic decay. We make the following observations: (i) sensors are subject to noise, (ii) change of the process

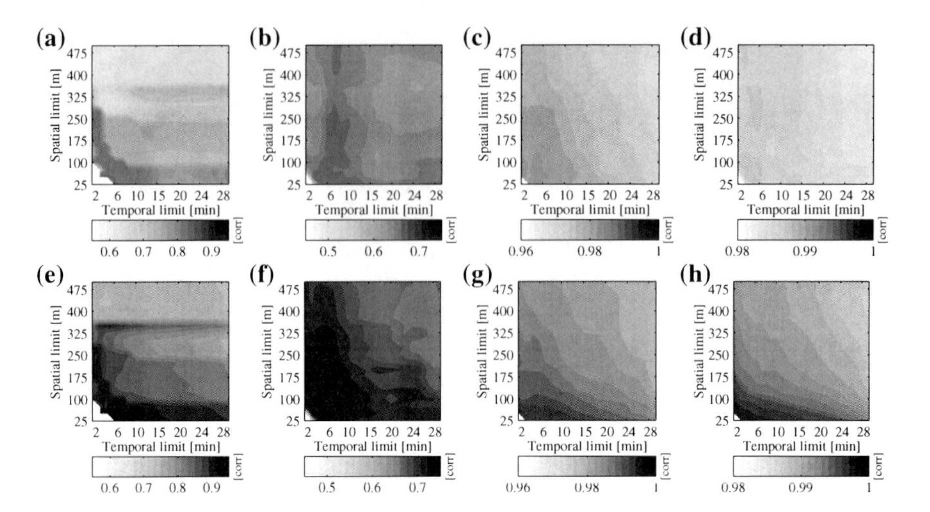

Fig. 3 Pearson correlation of the monitored phenomena for different spatial $\hat{d}$ and temporal $\hat{t}$ thresholds. **a** CO. **b** O_3. **c** Humidity. **d** Temperature. **e** CO [s]. **f** O_3 [s]. **g** Humidity [s]. **h** Temperature [s]. [s] denotes signal aggregation with window size of 20

might be too fast for our measurement schedule, and (iii) change of the process might be too local for our requirements on spatial closeness. We smooth the values in rendezvous pairs taken with the same sensor with a moving average to address the first two issues. Note that sparse sampling can be compensated by signal aggregation [34]. In the second row of Fig. 3, we smooth the sequences by using moving averages with window size of 20 before computing the correlation coefficients. For all considered phenomena we were able to obtain a more coherent picture of the decay of spatial and temporal relation between sequences of measurements.

Our results suggest that the variation of CO in the air is very local and requires spatial closeness of < 100 m (Fig. 3a, e) for two sensors making a rendezvous. Variation of humidity and temperature is minor over the considered temporal and spatial limits. This is confirmed by high correlation (> 0.95) of rendezvous pairs for all points in the plots (Fig. 3c, d, g, h). In this case, the improvement gained through data aggregation is insignificant due to low sensor noise as can be concluded from a very small increase of average correlations.

The bottom left corner in all subfigures of Fig. 3 does not contain any data as a result of low number of rendezvous pairs ($|\Phi| < 1{,}000$) and is thus white-colored. We observe an important trade-off between correlation of rendezvous pairs and their number. On the one hand, tighter rendezvous parameter values lead to a higher correlation of values. On the other hand, it takes longer to obtain a sufficient number of rendezvous pairs to enable their useful applications, such as sensor fault detection and sensor calibration.

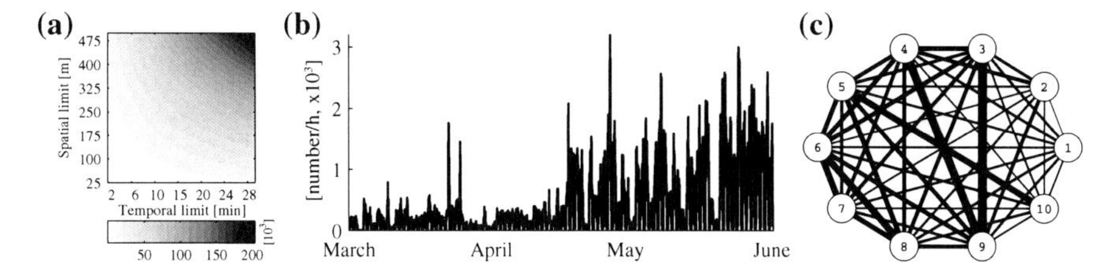

Fig. 4 Average number of rendezvous pairs between two sensors increases significantly with increasing spatial and temporal rendezvous parameters (**a**). The number of rendezvous pairs over time ($\hat{d} = 100\,\text{m}, \hat{t} = 10\,\text{min}$) depends on the number of nodes in operation (**b**) and is unevenly distributed between them (**c**). Thickness of lines is proportional to number of rendezvous pairs. **a** average, **b** over time, **c** pairwise

4.2 Rendezvous Connection Graph

The number of rendezvous pairs depends on the parameters $\hat{d}$ and $\hat{t}$. Analysis of the OpenSense dataset reveals that the number of rendezvous pairs increases significantly for larger $\hat{d}$ and $\hat{t}$, cf. Fig. 4a. On average, a pair of sensors located up to 100 m from each other and taking measurements up to 10 min apart makes over 20,000 parallel measurements over the period of three months. We plot the total number of rendezvous pairs per hour for this parameter setting in Fig. 4b. One streetcar was on maintenance before mid-April, which explains the 9x increase of rendezvous pairs over the last 1.5 months. The distribution of rendezvous pairs among nodes is uneven, as depicted in Fig. 4c (line thickness is proportional to the number of rendezvous pairs) and depends on whether two streetcars share the same depot and thus serve the same set of lines.

A large number of rendezvous pairs per se does not ensure that one can compare any two sensors in a mobile network. We introduce a *rendezvous connection graph* as an undirected graph with a set of sensors as its vertices, and a set of edges between sensors, which make a rendezvous. Connectivity of the rendezvous connection graph is crucial for identifying sensor failures and updating sensor calibration as will be shown in the next section. In Fig. 5, we plot the probability of having the OpenSense connectivity graph connected within a given time limit in days. We compare results

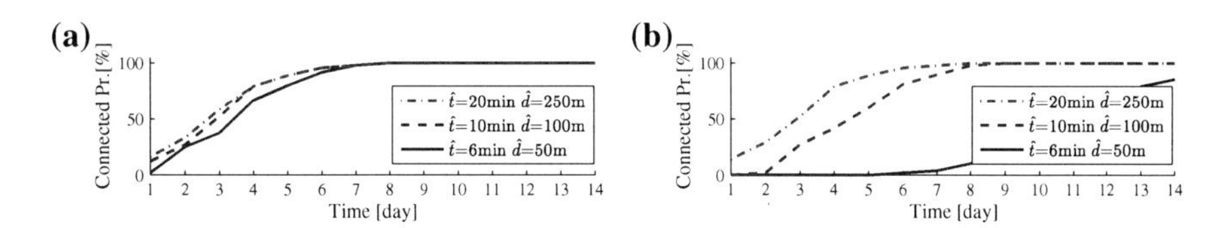

Fig. 5 Network connectivity analysis. Network connectivity depends on the spatial and temporal thresholds and on the required number of measurements per rendezvous. **a** Rendezvous with >1,000 measurements, **b** Rendezvous with >5,000 measurements

for cases when $|\Phi| \geq 1{,}000$ and $|\Phi| \geq 5{,}000$ rendezvous pairs are sufficient to qualify for a rendezvous. This is necessary in case of noisy sensors and gives the possibility to compute a correlation between rendezvous pairs. In all cases, our dataset suggests that within 4 days it is highly probable to connect the whole network. All nodes in the network are always reachable within 7 days. As can be seen in Fig. 5b, very tight rendezvous parameters may increase the time required for network connectivity, however, most processes monitored by the OpenSense network set more relaxed constraints on rendezvous (e.g., $\hat{d} = 100\,\text{m}$ and $\hat{t} = 10\,\text{min}$). Considering reference stations does not impact the time needed to reach all nodes in the network, indicating that a subset of OpenSense nodes passes by reference stations quite often.

5 Rendezvous Applications

We end the chapter with two case studies that use the introduced rendezvous to detect faulty sensors and calculate up-to-date calibration parameters for functional sensors. Based on our findings from Sect. 4 we fix the spatial and temporal threshold of a rendezvous to $\hat{d} = 100\,\text{m}$ and $\hat{t} = 10\,\text{min}$, respectively. For simplicity we use the same threshold for all environmental datasets.

5.1 Case Study: Faulty Sensor Detection

A faulty sensor often generates data which does not look conspicuous and does not conform to the typically used fault models [3]. This makes fault detection a non-trivial challenge. In this work, we consider a sensor *faulty* if its measurements do not correlate with the majority of mutually correlated sensors at rendezvous. A similar definition of a faulty sensor in static networks is used in [3]. Sensor faults might be a consequence of hardware failures (e.g., gas sensors can be destroyed if exposed to grease or organic solvents [30]) or sensor sensitivity loss and soaring noise due to sensor aging. Although our gas sensors are uncalibrated, they are not treated as faulty and give high correlations on account of linearity of their sensor output [30, 31]. Linear transformations have no effect on the Pearson correlation coefficient.

In case of easily-comprehensible environmental phenomena, such as temperature, it is often obvious whether a sensor is faulty or not. For example, the snapshot in Fig. 6a shows an obviously faulty temperature sensor considering that the sensor is calibrated and the measurements were taken in summer. A threshold filter would be able to detect this kind of faults. For many environmental phenomena like gases and even humidity, a faulty sensor might be difficult to recognize. Fig. 7a shows an example of a faulty humidity sensor, which is difficult to identify without further data.

Based on data from the OpenSense network we observe that a surprisingly low correlation of calibration pairs at a rendezvous indicates that one of the sensors

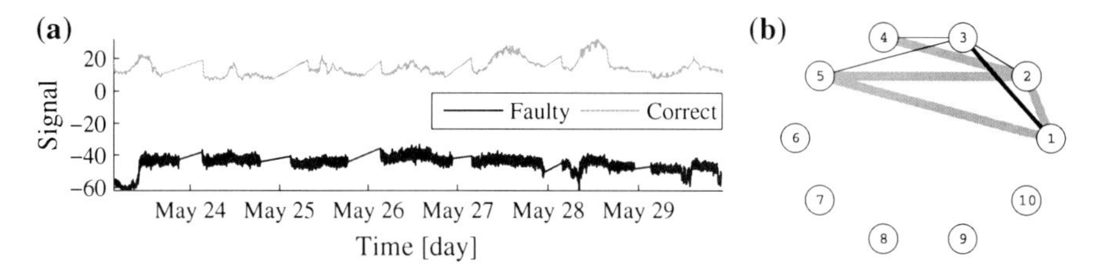

Fig. 6 The faulty temperature sensor of node 3 can be easily detected e.g., by a threshold filter. Correlation with other sensors is lower than between fully functional sensors. **a** Absolute measured values, **b** Pairwise correlation

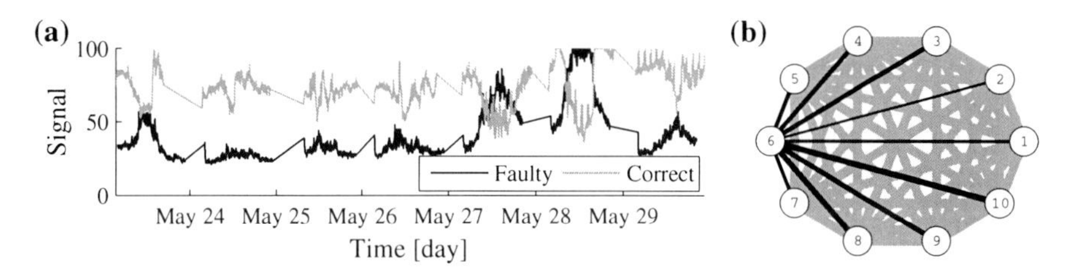

Fig. 7 The faulty humidity sensor of node 6 does not look conspicuous and does not conform to the typically used fault models. However, it reveals significantly lower correlation with other sensors when compared to the correlation of fully functional sensors. **a** Absolute measured values, **b** Pairwise correlation

is not working properly. We demonstrate this statement with the help of Fig. 6b and Fig. 7b, where both plots show pairwise correlations between temperature and humidity sensors attached to OpenSense nodes for $\hat{d} = 100\,\text{m}$ and $\hat{t} = 10\,\text{min}$. The thickness of the edge is proportional to the correlation of the nodes' measurements at a rendezvous. The failed temperature sensor is the internal temperature sensor built-in five OpenSense nodes. Only these OpenSense nodes have edges in the rendezvous connection graph in Fig. 6b. The average correlation between sensor 3 and other sensors is 0.32, whereas the average correlation between correctly working sensors is > 0.98. The faulty humidity sensor has an average correlation of 0.21 with other sensors, whereas the functional sensors have > 0.96 correlation between each other.

5.2 *Case Study: Sensor Calibration*

In the second case study we demonstrate how rendezvous can be used to calibrate low-cost sensors. The ten sensor nodes pass by two reference stations from time to time, which are located near the streetcar tracks. We make use of these rendezvous between sensor nodes and reference stations to calculate up-to-date calibration parameters for the ozone and temperature sensors deployed in the OpenSense nodes.

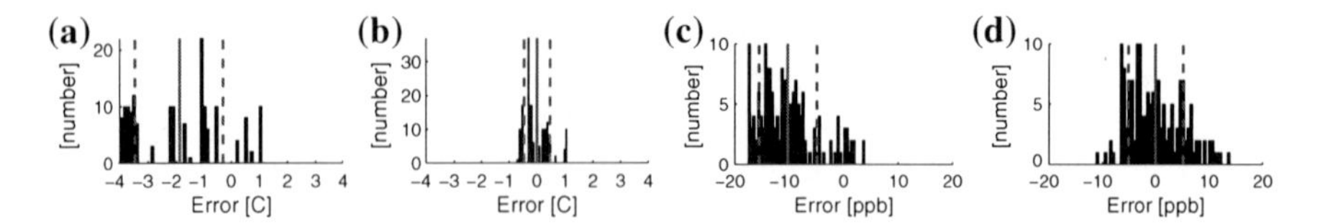

Fig. 8 Before and after using rendezvous to calibrate the temperature and ozone sensors of node 7. The colored lines denote the average *(straight)* and standard deviation *(dashed)* of the measurement errors when compared to measurements from the reference station. **a** Temp.: original, **b** Temp.: calibrated, **c** O_3: original, **d** O_3: calibrated

We illustrate the applicability of sensor calibration using rendezvous with a three-day data extract from sensor node 7. We use the method of least squares to adjust the calibration parameters of the sensors such that the sum of squared differences between sensor readings and reference measurements is minimized [35]. With this calibration approach, we are able to considerably increase the measurement accuracy of node 7, as depicted in Fig. 8. For temperature the mean absolute error is reduced from $2.1 \pm 1.6\,°C$ to $0.4 \pm 0.5\,°C$ and for ozone from 10.5 ± 5.3 ppb to 4.2 ± 5.1 ppb. Further, we could remove for both sensors an undesired bias as it can be seen on the mean error of 0 (colored straight lines in Fig. 8b, d).

6 Conclusions

This work introduced and explored the potential of phenomenon-based rendezvous between sensors in a mobile setting. For our analysis we used data gathered with the OpenSense network, which monitors air quality on top of streetcars in an urban area. Pairwise correlations between measurements of mobile sensors at various spatial and temporal distances revealed a dependency between the locality of a monitored phenomenon and the strength of correlation. This dependency impacts the parameters that define a rendezvous and control its duration and frequency. We presented the results for a dataset comprising CO, O_3, temperature, and humidity sensor readings. We found that all sensors in the OpenSense deployment can be reached within a few days to test the correct behavior of the whole network. As a case study of the developed concepts, we used rendezvous to detect sensor failures and to improve sensor calibration.

Acknowledgments We would like to thank Tonio Gsell and Jan Beutel for their technical support. Further, we thank Roman Lim, Federico Ferrari, and the anonymous reviewers for their valuable feedback that helped us to improve this chapter. This work was funded by NanoTera.ch with Swiss Confederation financing.

References

1. Beutel, J., Gruber, S., Hasler, A., Lim, R., Meier, A., Plessl, C., Talzi, I., Thiele, L., Tschudin, C., Woehrle, M., et al.: PermaDAQ: A scientific instrument for precision sensing and data recovery in environmental extremes. ACM/IEEE IPSN, In (2009)
2. Ceriotti, M., Mottola, L., Picco, G.P., Murphy, A.L., Guna, S., Corra, M., Pozzi, M., Zonta, D., Zanon, P.: Monitoring heritage buildings with wireless sensor networks: The Torre Aquila deployment. ACM/IEEE IPSN, In (2009)
3. Yim, S., Choi, Y.: Neighbor-based malicious node detection in wireless sensor networks. Wireless Sensor Network **4**(9), 219–225 (2012)
4. Farruggia, A., Re, G.L., Ortolani, M.: Detecting faulty wireless sensor nodes through stochastic classification. In: PerCom Workshops, IEEE. (2011) 148–153.
5. Miluzzo, E., Lane, N.D., Campbell, A.T., Olfati-Saber, R.: CaliBree: A self-calibration system for mobile sensor networks. IEEE DCOSS, In (2008)
6. Hasenfratz, D., Saukh, O., Thiele, L.: On-the-fly calibration of low-cost gas sensors. Springer EWSN, In (2012)
7. Rajasegarar, S., Leckie, C., Palaniswami, M.: Anomaly detection in wireless sensor networks. Wireless Communications, IEEE **15**(4), 34–40 (2008)
8. Hoek, G., Beelen, R., de Hoogh, K., Vienneau, D., Gulliver, J., Fischer, P., Briggs, D.: A review of land-use regression models to assess spatial variation of outdoor air pollution. Elsevier Atmospheric Environment, In (2008)
9. Elnahrawy, E., Nath, B.: Cleaning and querying noisy sensors. ACM WSNA, In (2003)
10. Hasenfratz, D., Saukh, O., Thiele, L.: Model-driven accuracy bounds for noisy sensor readings. IEEE DCOSS, In (2013)
11. Aberer, K., Sathe, S., Chakraborty, D., Martinoli, A., Barrenetxea, G., Faltings, B., Thiele, L.: OpenSense: Open community driven sensing of environment. ACM IWGS, In (2010)
12. Shi, K.: Semi-probabilistic routing in intermittently connected mobile ad hoc networks. Journal of Information Science and Engineering **26**(5), 1677–1693 (2010)
13. Xing, G., Wang, T., Xie, Z., Jia, W.: Rendezvous planning in wireless sensor networks with mobile elements. IEEE Transactions on Mobile Computing **7**(12), 1430–1443 (2008)
14. Park, J., Moon, K., Yoo, S., Lee, S.: Optimal stop points for data gathering in sensor networks with mobile sinks. Wireless Sensor Network **4**(1), 8–17 (2012)
15. Du, J., Liu, H., Shangguan, L., Mai, L., Wang, K., Li, S.: Rendezvous data collection using a mobile element in heterogeneous sensor networks. International Journal of Distributed Sensor Networks **12**, 1–12 (2012)
16. Choi, B.J., Liang, H., Shen, X.S., Zhuang, W.: DCS: distributed asynchronous clock synchronization in delay tolerant networks. IEEE Transactions on Parallel and Distributed Systems **23**(3), 491–504 (2012)
17. Cao, Q., Abdelzaher, T., He, T., Stankovic, J.: Towards optimal sleep scheduling in sensor networks for rare-event detection. ACM/IEEE IPSN, In (2005)
18. Li, J.J., Faltings, B., Saukh, O., Hasenfratz, D., Beutel, J.: Sensing the air we breathe-the OpenSense Zurich dataset. AAAI, In (2012)
19. Lee, C.H., Kwak, J., Eun, D.Y.: Characterizing link connectivity for opportunistic mobile networking: Does mobility suffice? In: INFOCOM. (2013) 2124–2132.
20. Schmid, S., Wattenhofer, R.: Algorithmic models for sensor networks. WPDRTS, In (2006)
21. Abrams, Z., Goel, A., Plotkin, S.: Set k-cover algorithms for energy efficient monitoring in wireless sensor networks. In: ACM/IEEE IPSN. (2004) 424–432.
22. Gui, C., Mohapatra, P.: Power conservation and quality of surveillance in target tracking sensor networks. In: ACM MobiCom. (2004) 129–143.
23. Hsin, C.f., Liu, M.: Network coverage using low duty-cycled sensors: random & coordinated sleep algorithms. In: ACM/IEEE IPSN. (2004) 433–442.
24. Kumar, S., Lai, T.H., Balogh, J.: On k-coverage in a mostly sleeping sensor network. In: ACM MobiCom. (2004) 144–158.

25. Cevher, V., McClellan, J.H.: Sensor array calibration via tracking with the extended kalman filter. In: ICASSP, IEEE Computer Society (2001) 2817–2820.
26. Girod, L., Lukac, M., Trifa, V., Estrin, D.: The design and implementation of a self-calibrating distributed acoustic sensing platform. In: ACM SenSys. (2006) 71–84.
27. Koushanfar, F., Univ, R., Taft, N., Potkonjak, M.: Sleeping coordination for comprehensive sensing using isotonic regression and domatic partitions. In: INFOCOM. (2006) 1–13.
28. Keally, M., Zhou, G., Xing, G., Wu, J.: Exploiting sensing diversity for confident sensing in wireless sensor networks. In: INFOCOM. (2011) 1719–1727.
29. Hwang, J., He, T., Kim, Y.: Exploring in-situ sensing irregularity in wireless sensor networks. In: ACM SenSys. (2007) 289–303.
30. SGX Sensortech: MiCS-OZ-47 ozone sensor. http://goo.gl/kZ5ay
31. Alphasense: CO-AF sensor on a digital transmitter board. http://goo.gl/9Bl5n
32. Sensirion: SHT10 humidity and temperature sensor. http://goo.gl/LHBas
33. Aberer, K., Hauswirth, M., Salehi, A.: A middleware for fast and flexible sensor network deployment. VLDB, In (2006)
34. Vucetic, S., Fiez, T., Obradovic, Z.: Examination of the influence of data aggregation and sampling density on spatial estimation. Water Resources Research **36**(12), 3721–3730 (2000)
35. Björck, A.: Numerical methods for least squares problems. SIAM, In (1996)

Real-Life Deployment of Bluetooth Scatternets for Wireless Sensor Networks

Michael Methfessel, Stefan Lange, Rolf Kraemer, Mario Zessack, Peter Kollermann and Steffen Peter

Abstract Bluetooth scatternets are constructed from overlapping piconets, allowing any number of nodes to be connected into a multi-hop wireless network. Although the topic has been researched for 15 years, no deployments of self-organized scatternets have been published. Recently we have presented the SFX algorithm, which was implemented on commercial Bluetooth nodes and is an extension of SHAPER from 2003. Here measurements are presented for scatternet trees for a laboratory network of 24 nodes and for deployment in a photovoltaic power plant with 39 nodes. The results demonstrate the effectiveness of the SFX algorithm, which is evidently the only actually implemented scatternet contruction procedure which is distributed and does not assume full node-to-node visibility.

1 Introduction and Related Work

Developed in the 1990's, Bluetooth [3] is widely used for short-distance point-to-point wireless links. The Bluetooth standard also introduced the concept of a scatternet. A scatternet is a collection of Bluetooth piconets, each containing up to eight nodes, forming a multi-hop network. Bluetooth is a cheap, robust, and verified technology. Therefore scatternets are good candidates for wireless sensor networks for industrial control and surveillance and similar systems.

M. Methfessel · S. Lange (✉) · R. Kraemer
IHP, Im Technologiepark 25, 15236 Frankfurt(Oder), Germany
e-mail: lange@ihp-microelectronics.com

M. Zessack
lesswire AG, Rudower Chaussee 30, 12489 Berlin, Germany

P. Kollermann
AE REFUsol GmbH, Uracher Strasse 91, 72555 Metzingen, Germany

S. Peter
Center for Embedded Computer Systems, University of California, Irvine, USA

K. Langendoen et al. (eds.), *Real-World Wireless Sensor Networks*,
Lecture Notes in Electrical Engineering 281, DOI: 10.1007/978-3-319-03071-5_4,

Interestingly, the Bluetooth standard left open the procedure by which a scatternet is set up. This has lead to intense research and a number of distinct proposals, in particular for the most useful case: a distributed, self-organizing and self-healing algorithm which does not assume full node-to-node visibility. However, to the best of our knowledge, none of these have made it to a working implementation on actual Bluetooth hardware. Consequently, although scatternets with a fixed predefined topology were used in some applications, self-organized scatternets have not been used in any practical applications to date.

The authors have recently presented a distributed algorithm to build a scatternet with the desired self-organizing, self-healing, and self-optimizing properties, which does not assume full node-to-node visibility. This SFX procedure is based on the previously presented SHAPER algorithm. The basic idea (to move the node of one tree to a suitable position by a series of role switches prior to merging, see Fig. 1) is also used in the SFX algorithm. However, while working on a real-world implementation of SHAPER, key points arose for which solutions were still required. These concern (1) the collection of information about neighboring nodes, (2) overall steering of the algorithm, (3) locking and (4) tree optimization. Solutions were found for these issues and presented in Ref. [7].

In this chapter, results for tree construction in a laboratory test network of 24 nodes are presented. Furthermore, as a realistic application the SFX algorithm was deployed in photovoltaic power plant, where it has been functioning reliably for several months. The purpose of the installation is to collect diagnostic data from the inverters without need for a cable-based Ethernet network. The measured data demonstrates that the SFX procedure indeed reliably builds, optimizes, and maintains a tree with the required properties.

Comparing to previously proposed scatternet formation algorithms [11], these are generally mesh-based [12] or tree-based, with some discussion of star and ring topologies [1]. A mesh offers redundancy which can quickly compensate for disrupted links, while a tree must be repaired to regain connectivity. On the other hand, a mesh requires extra effort for routing which is avoided in a tree. Real-world considerations argue against some of the proposed procedures. Centralized approaches

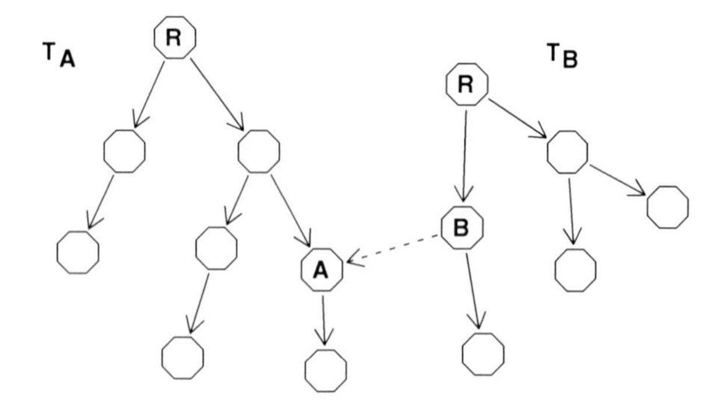

Fig. 1 The basic merge procedure as introduced by SHAPER. *Arrows* point from master to slave and *R* labels root nodes. Trees T_A and T_B will merge by building up a new connection (*dashed*) with node *B* as master and node *A* as slave. Prior to connecting, the root of T_A is moved to node *A* by a sequence of role switches

such as BTCP [9] or also MMPI [5] can construct an optimal piconet structure, but must collect and distribute information from a central site. When all nodes of the system can communicate with each other, elegant techniques such as TSF [10] or the closely similar TreeNet [2] can be used. However, full node-to-node visibility is not given for most actual systems.

The most realistic approach seems to be SHAPER [4], on which the present work is based. In the following, Sect. 2 briefly characterizes the extensions added by the SFX algorithm. Results for the laboratory test network and the photovoltaic power plant are presented in Sects. 2 and 3, respectively, while Sect. 4 contains the conclusions.

2 Main Points of the SFX Algorithm

Figure 1 illustrates the basic SHAPER procedure to merge two trees. A sequence of role switches is applied to the links between the left-handed root node and A. The result is that A becomes root and is not slave in any piconet. This node is therefore free to become slave in the new connection.

SFX introduced four extensions to obtain a procedure which works reliably and effectively for real-world Bluetooth systems (see [7] for details):

Collecting information about the neighborhood A central task is to acquire information about other nodes in the neighborhood. The usual Bluetooth procedure (used by SHAPER) is to find a node using inquiry, build a connection, and exchange information over this connection. This turns out to be prohibitively inefficient in practice. SFX packages the relevant information as user-defined data in an extended inquiry response. By this simple device, a node has a full overview of all neighboring nodes and the quality of links to them after each inquiry phase. Effectively, each node is broadcasting its local topology to all nodes in the neighborhood.

Steering the algorithm When two trees merge, SHAPER reconfigures the smaller tree. Thus each node must know the total number of nodes in its current tree. When multiple distributed merge operations are done, it is a major task to maintain the tree cardinality consistently. Reference [4] does not adress the issue but presents simulation results; perhaps these simulations were on a higher level which assumes that the cardinality is known *a priori*. In a real-life situation, this sub-task has a similar complexity as the whole tree-building procedure.

SFX uses the following approach to enable a distributed algorithm. First, by means of an asymmetric merge procedure with one "active" and one "passive" tree, any number of merges can be done into the same target tree simultaneously. Secondly, merge operations are only permitted if the active tree can lower its tree identifier by merging. This avoids loop formation by simultaneous antiparallel merges betwen two trees. Furthermore, when a node merges into the ever-growing final tree (with the globally lowest tree identifier) it will be passive in subsequent merges. Therefore the tree which reconfigures is usually the smaller tree.

Tree locking Antiparallel merging is avoided as described above, but it must still be ensured that each tree does only one active merge at a time. Reference [4] suggests a solution, which however requires modification in order to cope with delays and possible race conditions arising when actual Bluetooth nodes are used. Details can be found in [7].

Tree optimization To optimize the tradeoff between tree depth and link quality, SFX introduced two procedures, whereby links are compared according to the measure $X_k = P_k + \beta D_k$ where P_k is the path loss for a link to node k and D_k is depth the node would have when connected via this link. First, after each inquiry phase a node inspects possible links within the same tree and switches its position if X can reduced. Second, a whole subtree can disconnect, "flip" by switching master and slave roles betwen the old and new bridge nodes, and reconnect. This procedure is quite similar to basic tree merging and is required in cases where the simple optimization cannot improve the situation.

3 Laboratory Tests

As a test, 24 nodes were distributed over two wings of the IHP building (Fig. 2). The nodes consist of a Bluetooth Class 1 module [3] and a μC unit running a Linux OS (including the full TCP protocol stack) and the SFX algorithm. By means of

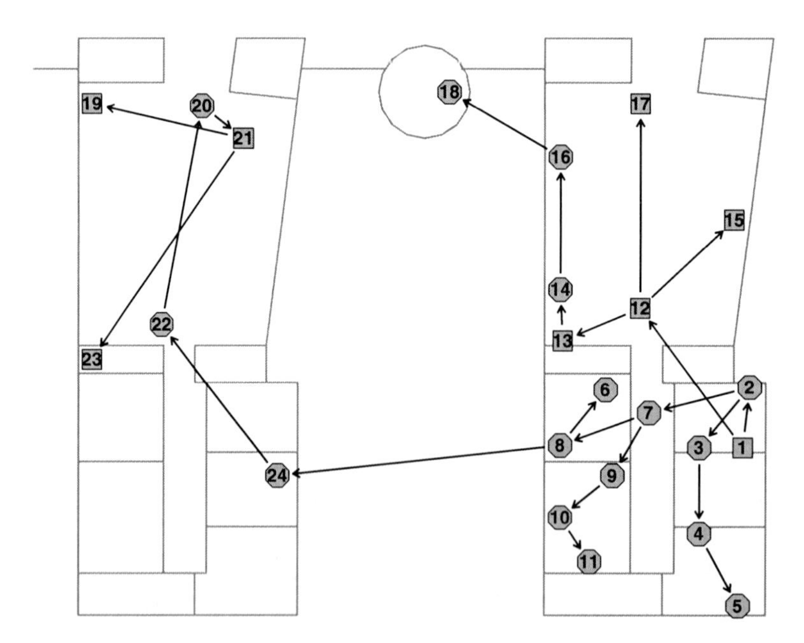

Fig. 2 Sketch of the in-house test network with 24 bluetooth nodes arranged in two adjacent wings (separation 20 m) over two floors (*squares above*, *circles below*). The *arrows* show a typical stable tree. The root is at node 1

a dedicated power amplifier and low noise amplifier, distances up to 500 m can be bridged with 6 cm dipole antennas [6]. All nodes were connected to an Ethernet backbone for easy control and monitoring. Nodes typically find 2 to 4 neighbors with a path loss below 85 dB within a wing. Links between the wings are weaker due to a metallic coating on external windows, with a path loss of ~100 dB from $N8$ to $N24$.

Repeated runs were done to study the performance during tree formation. The nodes booted at random in a ten-second interval, requiring another 18 s to boot. Starting with pairwise connections, nodes coalesce into trees. When a node joins the gateway tree containing $N1$, it optimizes its position within the tree according to the tradeoff between path loss and depth.

In the right-hand wing, most nodes can build robust connections to several nodes in their neighborhood. Therefore, the optimization procedure has enough degrees of freedom to reduce tree depth without sacrificing link quality. As a result, trees tend to look quite different in the right wing but the depth here is never greater than four. In the left-hand wing, links are weaker and the same local tree is constructed in almost all most cases. The link between the wings settles down to $N8 \rightarrow N24$ in the majority of cases.

To characterize the time required to set up the tree, Fig. 3 summarizes the time until all nodes have joined the gateway tree for 100 runs. The time includes the boot delay and lies at around 50–100 s. Histogram B shows the time until optimization is completed. Stability was defined as the start of the first 60-s period in which the tree remained static. The results show that two to three minutes are adequate to build up an optimized tree.

As described in Ref. [7], development was done in parallel on the actual hardware and in a simulation environment, both running the identical code. These simulations can now be used to study the behavior for larger systems. Nodes were spread randomly over a rectangular area, with parameters similar to those in the test network. The simulation results are shown in Fig. 4. We included the case for 24 nodes

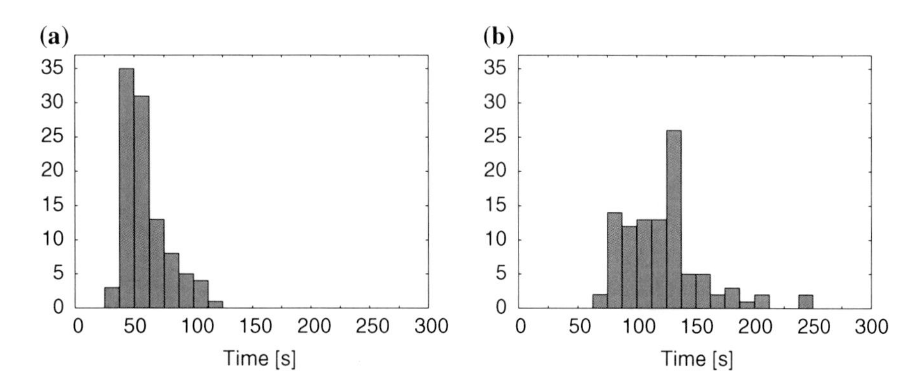

Fig. 3 Histogram showing the elapsed time until connectivity is reached (**a**) and until the tree is stable (**b**) for 100 runs of the 24-node IHP test network

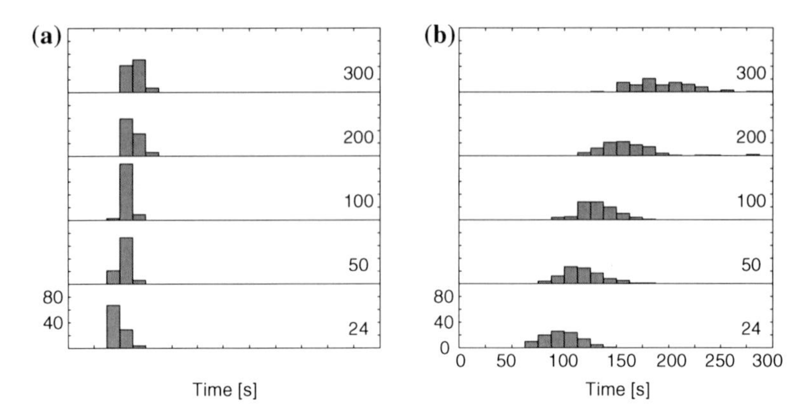

Fig. 4 Histogram showing the elapsed time until connectivity is reached (**a**) and until the tree is stable (**b**) for 100 simulation runs each for 24, 50, 100, 200, and 300 nodes

to compare with the measurements. Agreement is satisfactory with slightly shorter delays since the simulation uses a static channel model and disregarded the fact that inquiry responses are often corrupt in practice. The simulations show that the SFX algorithm works effectively for node counts in the hundreds. The times to reach connectivity and stability increase somewhat with the node count. The effect is larger for stability since optimization becomes more complex for large systems. In contrast, connectivity is still reached quickly, since merges are performed in all parts of the system in parallel.

4 Deployment in a Photovoltaic Power Plant

As a real-life test, 39 nodes were attached to the inverters of a photovoltaic power plant as shown in Fig. 5. The main issue was to test the effectiveness of the scatternet build-up and the robustness of data transfer under realistic conditions. Previous measurements had shown that radio transmission is made difficult by reflections from the rows of metal struts and from the need to avoid shadows on the photovoltaic modules. The nodes were mounted behind the upper edges of the modules. After finding poor results using small PCB-mounted ceramic antennas, good performance was obtained using 20 cm antennas which protruded above the edge. Information about the system can be found in Ref. [8].

A new tree is built up each morning as the nodes boot. Monitoring this process over 6 months, a correct tree was constructed in each case, except for several days when snow covered the solar modules. A typical generated tree is shown in Fig. 5. Usually the tree depth was 9 or 10 but depths down to 6 were also observed. Links over distances of 300 m or more occur. The pathloss for any link did not exceed 95 dB, the threshold to trigger subtree optimization. Thus, the algorithm is performing in

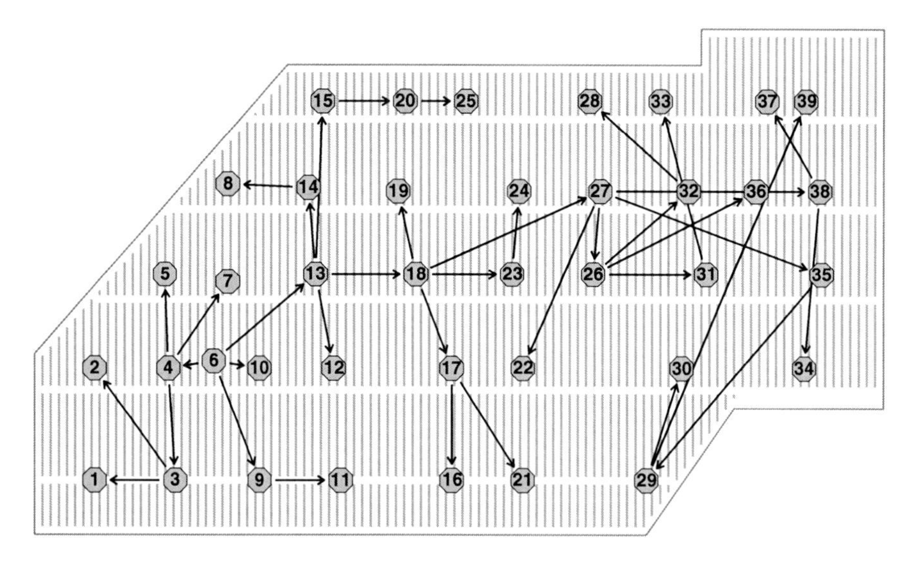

Fig. 5 Map of the photovoltaic power plant, showing the positions of the 39 nodes and a typical tree generated by the SFX procedure. The root is node 6. The field is 1300 by 550 m with 102 rows of solar modules (courtesy of GP JOULE)

accordance with the chosen parameter values. Typically, nodes boot within 2 min of each other. The tree is stable about one minute after the last node is activated. After that, optimization modifies the tree from time to time (Fig. 6) with a brief loss of connectivity. Since the nodes are located far apart, the algorithm was configured to search for optimal connections aggressively. Reconfigurations are triggered by changes in the transmission properties, e.g., due to weather conditions. In scenarios with shorter distances, parameters can be chosen to reduce the number of reconfigurations.

Measurements were done to characterize the data transfer in the scatternet. Figure 7 shows the ping times and the times required to transfer data via TCP. Both tests were done one node at a time, avoiding issues of competing data streams through

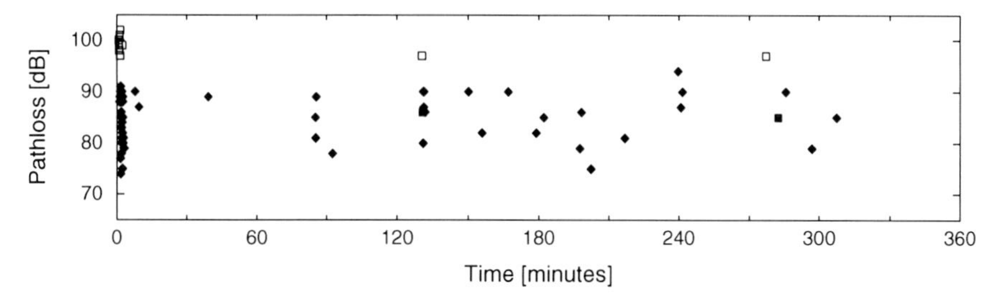

Fig. 6 Tree-maintenance events for 6 h after the scatternet boots (*diamonds* immediate optimization, *open squares* subtree optimization requests, *filled squares* subtree optimization). Values show the path loss associated with the event

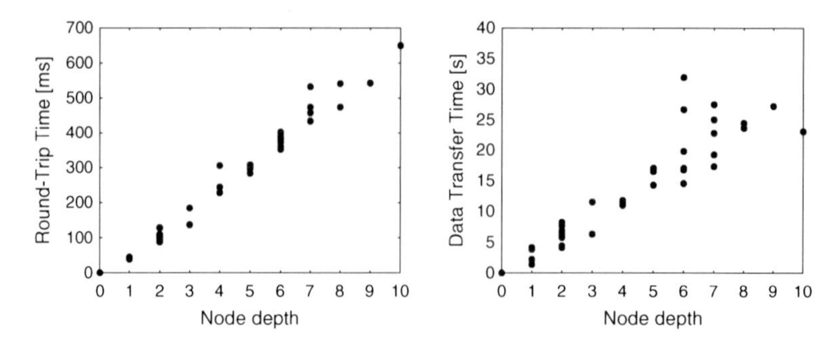

Fig. 7 Round-trip ping times (*left*) and time needed to transfer one Mbit of data (*right*) between the root and some other node for the power plant deployment

a bottleneck. The results are displayed as function of the depth in the tree of the target node. The evident linear behavior shows that good links were built up between all node pairs, which in the end is the main purpose of the procedure. The ping round-trip times show a delay of 30 ms per hop, as was also observed in the laboratory. A large part of the delay is due to the protocol handling in the Linux operating system and the Bluetooth stack.

5 Conclusions

Bluetooth scatternets have been the subject of research for many years. Ideally, a scatternet should set itself up by a distributed algorithm which does not assume full node-to-node visibility. The scatternet should automatically optimize and respond suitably when nodes enter or fail or when link properties change.

Of the previously proposed distributed algorithms for scatternet set-up and maintenance, none have made it to a deployable system. In this chapter, results were presented for the SFX algorithm, which constructs a scatternet tree in accordance with the requirements above and was implemented on commercially available Bluetooth modules. Starting from the basic idea of the SHAPER algorithm, extensions were added to obtain a real-life implementation. Simulations, laboratory tests, and deployment have demonstrated good performance:

- Tree build-up is fast, creating a stable and optimized tree in a few minutes.
- Measurements show a balanced tree with good links along all connections.
- Realistic simulations for hundreds of nodes demonstrate a good scalability.

To our knowledge, these successful results are the only study of self-organized large-scale Bluetooth scatternets in a realistic deployment to date.

Acknowledgments This work was supported by the German Bundesministerium für Bildung und Forschung (grants KF2123403DF9/KF2310001DF9). We thank GP JOULE for access to the photovoltaic power plant.

References

1. Al-Kassem, I., Sharafeddine, S., Dawy, Z.: Bluehrt: hybrid ring tree scatternet formation in bluetooth networks. In: IEEE Symposium on Computers and Communications, 2009. ISCC 2009, pp. 165–169 (2009)
2. Beutel, J.: Robust topology formation using btnodes. Comput. Commun. **28**(13), 1523–1530 (2005)
3. Bluetooth SIG.: Specification of the bluetooth system, v2.1+edr (2007)
4. Cuomo, F., Di Bacco, G., Melodia, T.: Shaper: a self-healing algorithm producing multi-hop bluetooth scatternets. In: IEEE Global Telecommunications Conference, 2003. GLOBECOM '03, vol. 1, pp. 236–240 (2003)
5. Donegan, B.J., Doolan, D.C., Tabirca, S.: Mobile message passing using a scatternet framework. Int. J. Comput. Commun. Control **III**(1), 51–59 (2008)
6. Lesswire AG, Berlin: BlueBear—industrial long range HCI-Bluetooth 2.0 Modul with EDR (2009)
7. Methfessel, M., Peter, S., Lange, S.: Bluetooth scatternet tree formation for wireless sensor networks. In: 2011 IEEE 8th International Conference on Mobile Adhoc and Sensor Systems (MASS), pp. 789–794 (2011)
8. Refuconnect.: REFUsol. http://www.refusol.com/accessories/refuconnect/ (2012)
9. Salonidis, T., Bhagwat, P., Tassiulas, L., LaMaire, R.: Distributed topology construction of bluetooth wireless personal area networks. IEEE J. Sel. Areas Commun. **23**(3), 633–643 (2005)
10. Tan, G., Miu, A., Guttag, J., Balakrishnan, H.: Forming scatternets from bluetooth personal area networks. Technical Report MIT-LCS-TR-826, Massachusetts Institute of Techonology (2001)
11. Whitaker, R.M., Hodge, L., Chlamtac, I.: Bluetooth scatternet formation: a survey. Ad Hoc Netw. **3**(4), 403–450 (2005)
12. Yu, C.M., Lin, J.H.: Enhanced bluetree: a mesh topology approach forming bluetooth scatternet. IET Wireless Sens. Syst. **2**(4), 409–415 (2012)

Part II
Poster and Demo Abstracts

Poster Abstract: Velux-Lab—Monitoring a Nearly Zero Energy Building

Alessandro Sivieri

Abstract Velux-Lab is the first Italian Nearly Zero Energy Building (NZEB) built inside a university campus; its main characteristic is the high energy efficiency. It constitutes an experimental module conceived as a laboratory to test new technologies and materials for energy efficiency in construction building. A Wireless Sensor Network collects data from several sensors monitoring the building and the HVAC system, and these data have been used to validate some properties of the building itself.

1 Introduction

Politecnico di Milano is the first Italian university to have a Nearly Zero Energy Building (NZEB) inside one of its campuses, the Velux-Lab (Fig. 1a).

This building is a project originally conceived for the Velux company as the model home for Mediterranean climate, designed by J.A. Cantalejo and A. Ronda; it has been shown in several locations around Europe (Bilbao, Rome, Milan) since 2007. In 2010, Velux decided to donate the building to the university, to be used as an energy laboratory to show and demonstrate new materials and technologies for energy efficiency.

The current building has the same structure of the original one, but materials and insulation have been replaced, due to the age of the previous ones, damaged by the installation and dis-assemblage of the various parts to move the building from site to site. An adaptation to the Milan climate has also been made, in particular by optimizing the shape for the hot season and weather stress: the roof helps to achieve natural ventilation, maximum solar gain in the winter and shadows in summer. Due

A. Sivieri (✉)
Dipartimento di Elettronica, Informazione e Bioingegneria, Politecnico di Milano,
Piazza L. da Vinci 32, Milano, Italy
e-mail: sivieri@elet.polimi.it

K. Langendoen et al. (eds.), *Real-World Wireless Sensor Networks*,
Lecture Notes in Electrical Engineering 281, DOI: 10.1007/978-3-319-03071-5_5,

Fig. 1 Motes hardware and locations

to these facts, the building reacts instantly to the climate change and only in the extreme seasons it needs help from the HVAC system to maintain comfort for its users [1].

A monitoring system has been deployed inside the building, to collect data from several sources and help to validate the theoretical properties of the building itself; in particular:

- 14 temperature probes have been installed inside the different layers of the walls, to closely capture all the variations due to the different materials that constitute these layers; these probes are connected to 5 Tmote Sky devices through an expansion board specifically designed to perform high precision measurements
- the electrical meters located in the HVAC system are connected to a sink collecting the electrical consumption of several components of the system, to closely monitor the real efficiency of the building
- Co_2 sensors, ultrasonic meters and inverters paired with solar panels are being deployed in the building and will be integrated in the data collection in the next few weeks.

Engineers and architects from the university are then able to see the data from all these sensors, aggregate it and compare it to the theoretical properties of the building and the system, to check the real energy efficiency of the Velux-Lab.

The rest of the article will focus on the Wireless Sensor Network and data analysis described in the first item in the list shown above.

2 The Wireless Sensor Network

The Wireless Sensor Network is built using 5 Tmote Sky devices, located in one of the two rooms of the building, plus the sink connected to a computer in the HVAC room. Each device has 1–4 temperature probes connected to it through an expansion board conceived for high precision measurements.

The hardware The requirements for temperature measurements coming from the energy experts is to have a max uncertainty of 0.1 °C per probe (Fig. 1b), to be able to acquire the smallest variations in the differences between the wall layers. The ADCs of the motes we used do not have sufficient precision to satisfy these requirements, especially because the PT1000 probes we used have variations of less than 4 Ω for each degree, or less than 0.4 Ω for each tenth of a degree, so signal amplification is needed to take these small variations in resistance (thus in voltage) and digitally convert it with the needed precision. This expansion board (Fig. 1c) has a separate power supply, also on battery, because it needs more than the 3V provided by the mote to amplify the signal; the 4 AA batteries needed by the shield last more or less one year, with measurements taken every minute by all the connected probes.

The motes, powered by 2 AA batteries, are not able to last as long as the expansion board, for several reasons: first, the radio power cannot be maintained low, due to the fact that two of the five devices are located in cavities in the outside wall, where the temperature probes of the outer layers are located; the peculiar materials of these layers and the protection needed inside the cavities to avoid problems with averse weather conditions (rain and snow) require a significant power for the radio system. Another factor of the shortened battery duration is the sampling frequency: each mote sends once per minute three different kinds of data (described below), in a multi-hop network, causing collisions and retransmission of the packets. We decided to stick with the AA batteries due to size constraints inside the external boxes, and with this setup motes last 2–3 months.

The software The motes run the TinyOS operating system [2], and the network routing protocol in use is CCBR [3]: we chose this protocol mainly for two reasons: the first one is that it has been designed "in house", and this was an interesting real application to try it out, and the second being that, despite the fact that our scenario is static and not mobile, which was one of the guidelines in designing CCBR, it can be useful to have support for context-based messages: other than temperature probes, all motes collect data about the residual energy on the batteries and the values of internal sensors (e.g., to detect if humidity and maybe water is inside the external cavities), all with different timings, and we wanted to give our users the possibility of setting timers depending on both the type of data collected and the location of the nodes.

The sink saves all the data on a MySQL database, to be accessed and analyzed by the experts.

3 Data Analysis

An example of the analysis performed by the engineers and architects, with the help of the data collected from the sensors, is about the internal thermal comfort of the building [4]. Figure 2 shows a chart of the various temperature ranges defining different comfort zones (as specified by the International norm [5] and the Italian legislation [6]).

P1 and P2 are two of the internal probes, one located on the West wall and the other on the East windows; measurements were taken from October 2012 to April 2013, and the chart shows how class A satisfaction comprises more than 95 % values for P1, showing the resistivity of the wall layers, while it comprises less than 40 % values

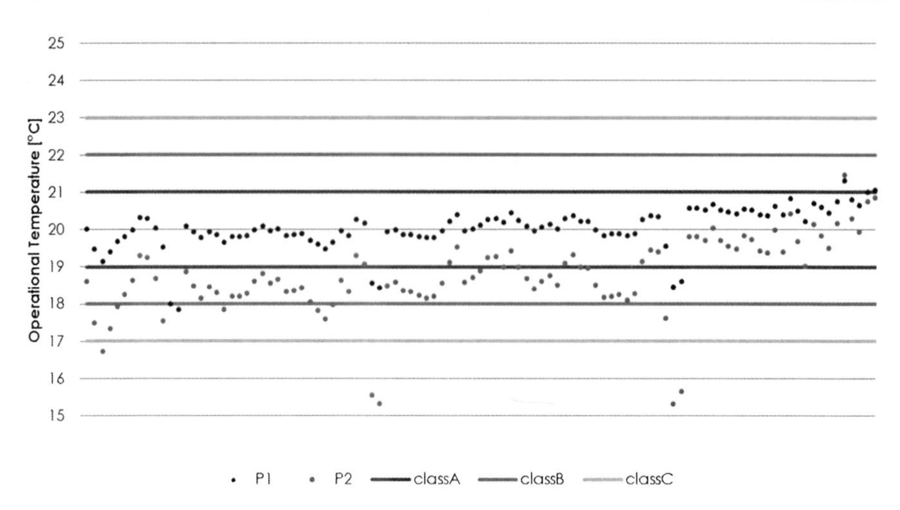

Fig. 2 Comfort zone measured by the probes

for P2, which disagrees with the theorical results obtained by dynamic simulations[1]: this shows on the one hand the limitations of the simulation process, and on the other hand the criticity of doors and windows in highly insulated buildings.

4 Conclusions and Future Work

The Wireless Sensor Network built to monitor this particular building, albeit simple from a software point of view, required to overcome some hardware challenges, due do the precision requirements by the stakeholders involved; its usefulness has been proved by the data analysis excerpt described above.

The future work for the whole project involves the extension of the monitoring parts of the system to all the possible dimensions that can be useful to verify the building energy properties; this may involve further extending the network with new nodes with different hardware requirements to interface with the existing sensors. The engineers and architects involved in the project are also working on configurations that may involve introducing actuators for the HVAC system.

Acknowledgments We would like to thank Velux and Politecnico di Milano, in particular Prof. M. Imperadori and his collaborators from the BEST Department and Prof. M. Motta and his collaborators from the Energy Department, for the involvement of our research group into this project.

[1] Not shown here due to space constraints: they will be shown on the poster.

References

1. Activehouse-veluxlab. http://www.activehouse.info/cases/veluxlab
2. Hill, J., Szewczyk, R., Woo, A., Hollar, S., Culler, D., Pister, K.: System architecture directions for networked sensors. SIGPLAN Not. **35**(11), 93–104 (2000)
3. Cugola, G., Migliavacca, M.: A context and content-based routing protocol for mobile sensor networks. In: Proceedings of 6th EWSN Conference (2009)
4. Sauchelli, M.: Prove dinamiche—dynamic tests. Domus (2012)
5. UNI EN ISO certifications n. 7726–7730 (2002–2006)
6. D.P.R. 26 agosto 1993 (Italian legislation administrative decree), n. 412 (1993)

Poster Abstract: Visualization and Monitoring Tool for Sensor Devices

Lubomir Mraz and Milan Simek

Abstract The objective of this chapter is to present SensMap, a novel web-based interactive visualization, monitoring and managing tool which brings major advantages when compared with similar available solution. SensMap is built on the top of Xively web-cloud and it is not tied with any hardware platform. SensMap provides rich set of features, from real-time status observation of sensor devices to comprehensive historical data comparison among multiple devices.

1 Introduction

The visualization of sensors location and interpretation of their data has crucial importance, it gives the data their meaning. At least three factors should be visualized: (i) what was measured, (ii) when it was measured and (iii) where the origin of the data is located. The main objective of the SensMap [1] is to achieve these tasks in the most intuitive and straightforward way. Our research team known as Wislab (Wireless System Laboratory of Brno) has recently developed the framework for visualization of sensor devices and their data stored in a Xively cloud [2]. The SensMap [1] is written in JavaScript/HTML5 and Adobe Flex and it is built as framework of the web-based applications. It offers exploration of nodes in Outdoor perspective (using Google maps), Indoor perspective (visualization of nodes placed inside of building) and Topology perspective (showing radio links and its quality among nodes in one wireless network). In this chapter we introduce how to get sensor data to the Xively and mainly how to visualize them in the SenMap framework. The rest of the chapter is structured as follows: In Sect. 2 the related applications are

L. Mraz (✉) · M. Simek
Department of Telecommunicatons, WISLAB Laboratory, Brno University of Technology, Brno, Czech Republic
e-mail: mraz@phd.feec.vutbr.cz
URL: www.wislab.cz

K. Langendoen et al. (eds.), *Real-World Wireless Sensor Networks*,
Lecture Notes in Electrical Engineering 281, DOI: 10.1007/978-3-319-03071-5_6,

discussed. Subsequently in Sect. 3 we briefly outline the primer features of SensMap. In Sect. 4, the individual visualization perspectives of the framework are presented. The chapter is concluded in Sect. 5, where our future work is outlined.

2 Related Work

Nowadays, there are dozens of visualization applications dedicated to WSN. A great review of most of them can be found in literature [2–4]. One of the most famous WSN platforms comes from Memsic Company. Their system is equipped with Mote-View visualization framework [5]. Another visualization framework are TinyViz for TOSSIM [6], SpyGlass [7] or NetViewer [8]. All the mentioned visualization applications are highly promising tools for the sensor data visualization and network control. However, most of them are running only as the desktop application and they are often very closely coupled with the specific hardware or simulation environment. The SensMap framework outperforms these limitations. It runs as web-based application and it allows the visualization of the arbitrary sensing system over the world. Furthermore, it goes deeper and provides very detailed visualization of the sensors deployed not only in approximate area but also in the specific building and even the room.

3 Main Features of SensMap

The SensMap disposes of three visualization perspectives [1]. Each perspective is implemented as the stand-alone web application and can run in a separate browser tab. The Outdoor View Perspective shows physical location of sensor nodes in the geographical view by using Google maps. Indoor View perspective shows physical deployment of sensor nodes inside the building. Lastly the Topology View perspective allows visualization of logical topology of particular wireless sensor network based on LQI (link quality indicator). Generally, the main features of the SenseMap can be listed as follows. (i) Visualizes sensor devices and their data in real-time or within defined interval in the history, (ii) Visualizes status (health) of sensor devices, (iii) Visualizes sensor devices geographically and also inside of the buildings, (iv) Allows to add, remove and manage the sensor devices and buildings, (v) Shows graphs and compares values from different data sources, (vi) Exports data for further analysis, (vii) Searches the sensor devices in fulltext mode and also within the defined geographical region, (viii) Shows topology of wireless sensor networks and link quality between nodes.

In the current version of the SenseMap framework, data from arbitrary sensing platform need to be stored in the Xively public cloud. Xively's data exchange (Xively REST API) is based on REST architecture, where data transfer is realized over the HTTP protocol. The data are visualized to the users via the SensMap, see Fig. 1.

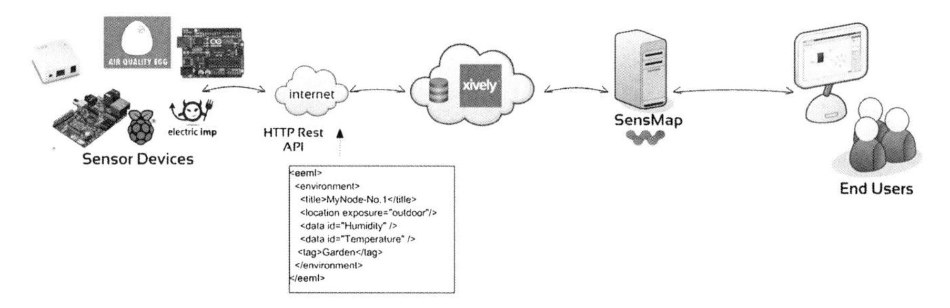

Fig. 1 The communication chain between user's data, Xively and SensMap

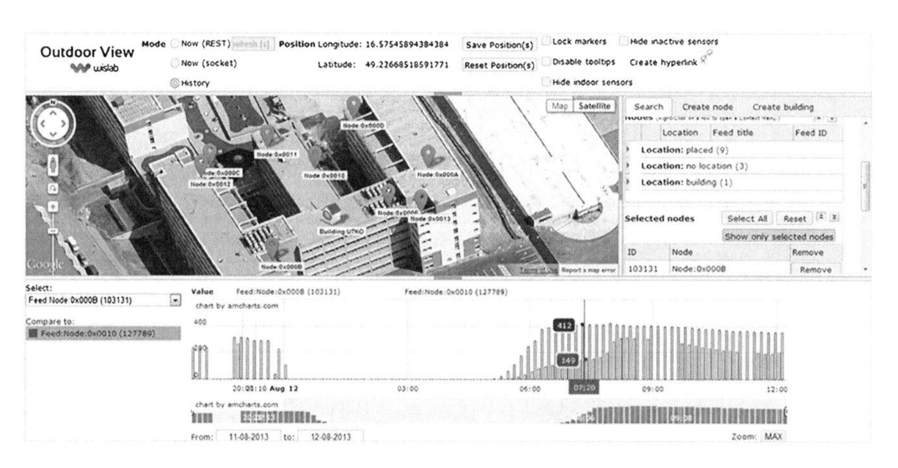

Fig. 2 OutdoorView perspective of SensMap

4 Visualization Perspectives

The SensMap framework is available at Ref. [1], where the OutdoorView perspective can be launched. At the OutdoorView link provided, our Wislab testbed with nine experimental sensor nodes can be found as it is depicted in Fig. 2. Each of the Wislab node disposes of four analog and two digital inputs. The measured data on individual inputs can be easily visualized. Each sensor device is graphically represented by the marker. Two types of markers can be distinguished within the application (i) the sensor marker and (ii) the building marker.

If the data from the sensor are not older than 15 min, the sensor marker is blue, otherwise it is grey. In the default state marker's label contains feed title. The user might select a datastream to be displayed in the label by right click on the node marker. This invokes a dropdown list which contains all the datastreams for the given node. Immediately after the selection a current value appears in the marker's label and from now, it is automatically updated (this is applied only for Now mode). The significant advantage of the Visualization Framework is presented by the simple management of the sensor devices through the graphical interface. Through the application, the new sensors can be added, placed or removed. After setting the Master Key in Outdoorview perspective, the user can easily add the new sensors into the Xively cloud . The operation with the IndoorView perspective is comparable with the OutdoorView. In addition, the user can visualize the network in the individual building floors and scale the size of the markers in order to fit them to the floor plan. The SensMap framework allows furthermore to observe the individual radio links and its quality between the sensors. For this purpose the SensMap offers the third perspective called TopologyView perspective. The user can see the network topology, where each sensor node is represented by a circle marker.

5 Conclusion

In this chapter, we presented SensMap, a flexible and modular web-based application for interactive visualization and monitoring of remote sensor devices. SensMap is used intensively in our research projects. Currently released v0.2 is still in the alpha stage. Beta release is expected in the end of 2013. As next steps, we would like to extend support for mobile sensor devices and implement an interface for triggering events over the data. SensMap is freely available and we believe that it can be useful for large sensing community.

Acknowledgments This chapter was prepared within the project SIX No. CZ.1.05/2.1.00/03.0072.

References

1. Wislab: SensMap-visualization and monitoring framework. Wislab. http://www.wsnapp.wislab.cz/
2. Xively: Xively-IoT public cloud. https://xively.com/ (2013)
3. Parbat, B., Dwivedi, A.K., Vyas, O.P: Data visualization tools for WSNs: a glimpse. Int. J Compute. Appl. **2**(1), 2010
4. Rodriguez Peralta, L.M.; Ieixeira Gouveia, B.A., Gomes de Sousa, D.J., da Silva Alves, C.: Enabling museum's environmental monitorization based on low-cost WSNs. In: New Technologies of Distributed Systems (NOTERE), pp. 227–234
5. Turon, M.: MOTE-VIEW: a sensor network monitoring and management tool. In: The 2nd IEEE Workshop on Embedded Networked Sensors (2005)
6. TOSSIM: Visualizing the real world. http://webs.cs.berkeley.edu/retreat-1-03/slides/1-03-tossim-tinyviz.pdf
7. Buschmann, C., Pfisterer, D., Fischer, S., Fekete, S.P., Kröller, A.: SpyGlass: a wireless sensor network visualizer. ACM SIGBED Rev. **2**(1), 1–6 (2005)
8. Ma, L., Wang, L., Shu, L., Zhao, J., Li, S., Yuan, Z., Ding, N.: NetViewer: a universal visualization tool for wireless sensor networks. In: Global Telecommunications Conference. IEEE GLOBECOM pp. 1, 5, 6–10 (2010). http://ieeexplore.ieee.org/stamp/stamp.jsp?tp=&arnumber=5683876&isnumber=5683069

Demo Abstract: MakeSense—Managing Reproducible WSNs Experiments

Rémy Léone, Jérémie Leguay, Paolo Medagliani and Claude Chaudet

Abstract Wireless Sensor Networks (WSN) users often use simulation campaigns before real deployment to evaluate performance and to fine-tune application and network parameters. This process requires repeating the same experiments under similar conditions and to collect, parse and present data efficiently. This chapter introduces MakeSense: a tool that automates this workflow and that allows reproducing simulations easily by defining the whole experiment and post-processing steps in a single JSON configuration file, easy to share and to modify. MakeSense also provides interfaces to interact with a running simulation, allowing to send external stimuli and to collect data in real time. MakeSense currently runs over the COOJA simulator, but has been built to be easily adapted to other architectures, including real testbeds.

1 Introduction

Evaluating applications and protocols through simulation always follow the same workflow from the simulation parameters definition to the creation of graphs that represent performance under different conditions. Yet, no generic tool really provides a way to automate the whole process and people often rely on ad-hoc scripts. Some

R. Léone (✉) · J. Leguay · P. Medagliani
Thales Communications & Security, Gennevilliers, France
e-mail: remy.leone@telecom-paristech.fr; remy.leone@thalesgroup.com

J. Leguay
e-mail: Jeremie.Leguay@thalesgroup.com

P. Medagliani
e-mail: Paolo.Medagliani@thalesgroup.com

R. Léone · C. Chaudet
Institut Mines-Télécom, Télécom ParisTech, CNRS LTCI UMR 5141,
Paris, France
e-mail: Claude.Chaudet@telecom-paristech.fr

K. Langendoen et al. (eds.), *Real-World Wireless Sensor Networks*,
Lecture Notes in Electrical Engineering 281, DOI: 10.1007/978-3-319-03071-5_7,

tools such as NEPI [3] focus on the interaction between testbeds, but do not address wireless sensor network platforms and hence, their architecture may be too complex for constrained devices.

In this chapter, we introduce MakeSense, a tool that automates large-scale WSN experiments and facilitates adaptation and reproducibility. MakeSense relies on a single JSON configuration file described in Sect. 2 from scenario definition to graphs generation. Sharing this lightweight file with the specific source code is sufficient to let others reproduce an experiment, improving the results trustworthiness [1]. MakeSense runs today over the COOJA simulator, but is designed to be generic and to be adapted to real testbeds too. We then describe its workflow and functionalities, including online interaction with simulation and describe a demonstration in Sect. 3.

2 MakeSense Core Functionalities

2.1 JSON Configuration File

All simulation parameters such as random seeds, transmission range are defined in a single JavaScript Object Notation configuration file. An excerpt of such a configuration file is presented on Listing 1.1. The file starts by the definition of some general parameters, such as the wireless interference range and specifies a template for the Makefile that will serve during the building process. It then defines one or several mote types (e.g. routers, end nodes, ...), the firmware they will use and gives them an RPL instance ID. In the `motes` section, it instantiates the different nodes with respect to the templates and specifies individual parameters such as their coordinates, their ID (address) and the UDP port on which they will be reachable during the simulation.

As this configuration file specifies all relevant parameters for a simulation, it is easy to use templates and scripting to run large experimental campaigns. A script can generate a series of such files, referring to the same or to different source codes, run experiments and store the JSON files as scenario descriptors, as they are lightweight.
Listing 7.1 Configuration file excerpt

```
[...]
"interference_range": 50.0,
"makefile_template":
    "simple_makefile",
"mote_types": {
  "hello": {
    "color": "blue",
    "description": "Router",
    "firmware_address": [
      "nodes",
      "hello",
      "hello-world.sky"
    ],
    "rpl_instance_id": 30
  }
},
"motes": [
  {
    "mote_id": 1,
    "mote_type": "hello",
    "settings": {
      "lambda": 1
    },
    "socket_port": 60001,
    "x": 10,
    "y": 0
},
[...]
```

2.2 *Workflow*

MakeSense workflow, represented schematically in Fig. 1, is composed of 6 steps that can be called independently, thanks to their loose coupling, or run in sequence using the `run_all` shortcut.

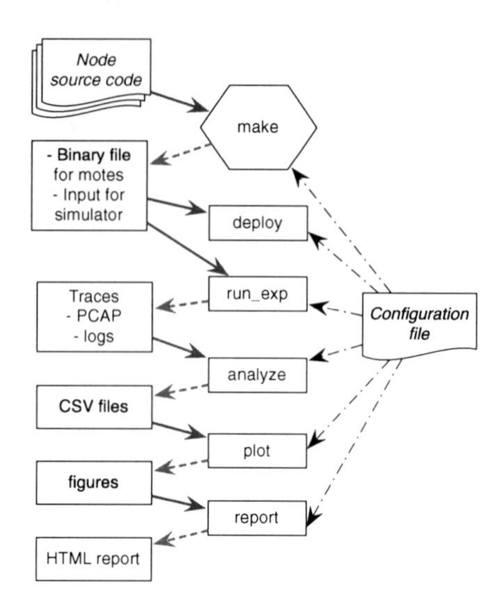

Fig. 1 MakeSense workflow

make creates the whole environment necessary for the experiment execution. It compiles the source code and creates the configuration files required by the simulation tool from templates.

deploy uploads the simulation files from a local repository to a remote location to run several simulations in parallel. This step can easily be adapted to upload firmwares to a real testbed.

run_exp launches the simulation series. MakeSense can launch concurrently, several functions that will fetch the simulation output through the nodes' serial or network interfaces. Each node has its own independent log file.

analyze parses trace files, e.g. PCAP files or text logs file, by applying a set of filters to produce CSV files that contain only the desired information. Filters are easy to specify and MakeSense includes a set of basic filters.

plot produces graphs using the CSV files produced by the analyze step and the info in the settings files to select relevant data.

report gathers all the results into a single HTML report file.

2.3 Multiple Control Channels

The simulation runs from its beginning to its end without user interaction, or can be run in real-time mode, allowing real-time interaction with traffic generators such as real applications, other simulators, or a user using an interactive command line and visualizing performance evolution through the COOJA scripting engine, as illustrated in Fig. 2. This feature allows to let the WSN run as if it were a real network and to measure its reaction to external, controlled, stimuli.

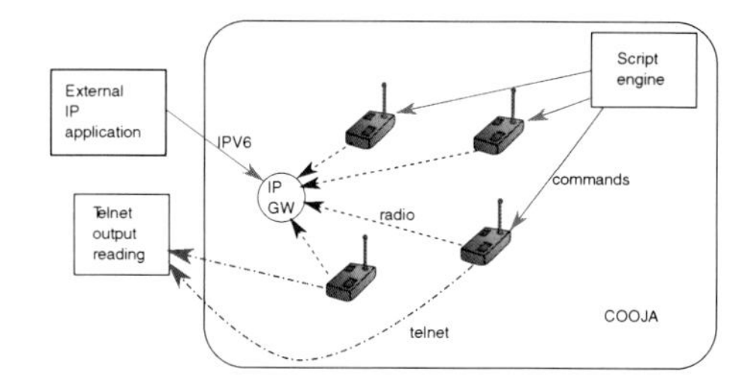

Fig. 2 Real-time interaction with COOJA

3 Demonstration

The COOJA implementation of MakeSense uses the Python programming language and other libraries such as fabric, jinja2 and matplotlib. As a demonstration, we use COOJA [2] in real time mode to simulate a network of 10 nodes connected using an RPL tree. The first node is a border router that connects all the remaining nodes that are CoAP servers, to the hosting operating system. We send traffic to the different sensor nodes from the shell and from the script engine contained inside the simulator. This traffic is composed of CoAP and PING requests.

As specified in the configuration file, MakeSense generates without user interaction various graphs. For example, Fig. 3 represents the theoretical connectivity graph, extracted from the sole configuration file using circular transmission range. Figure 4 is a representation of the real RPL tree, generated by querying nodes for RPL information on their serial interface, without perturbing the network traffic. Figure 5 is a classical per-protocol throughput graph, showing the use of per-protocols filters and Fig. 6 represents delay of Ping requests issued from the hosting operating system, that shows the interaction with the outside world.

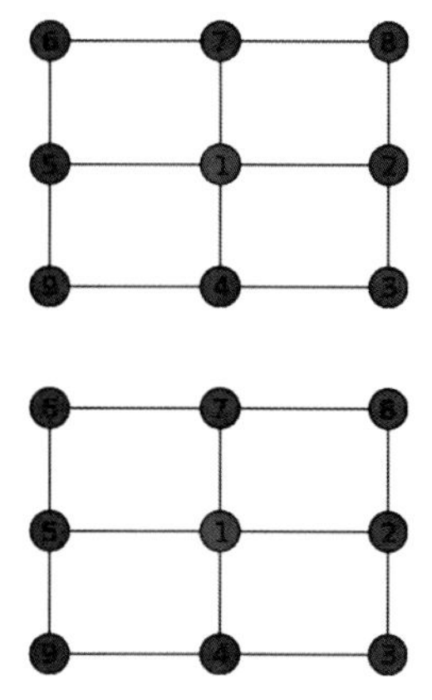

Fig. 3 Connectivity graph

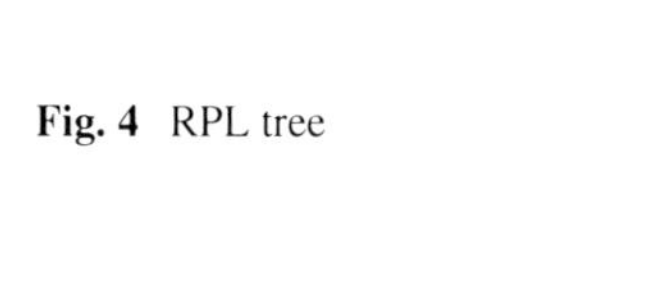

Fig. 4 RPL tree

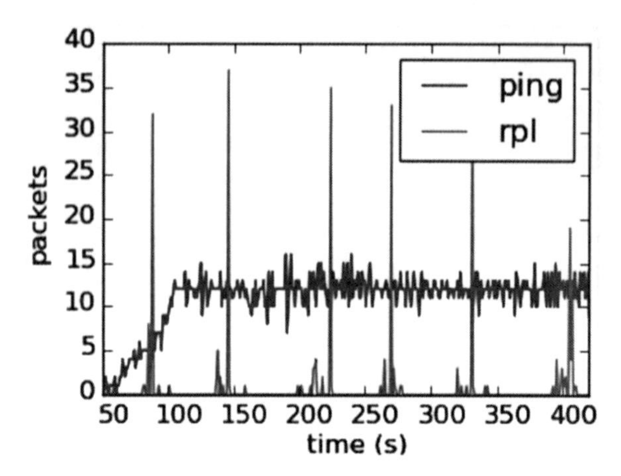

Fig. 5 RPL and ping traffic evolution

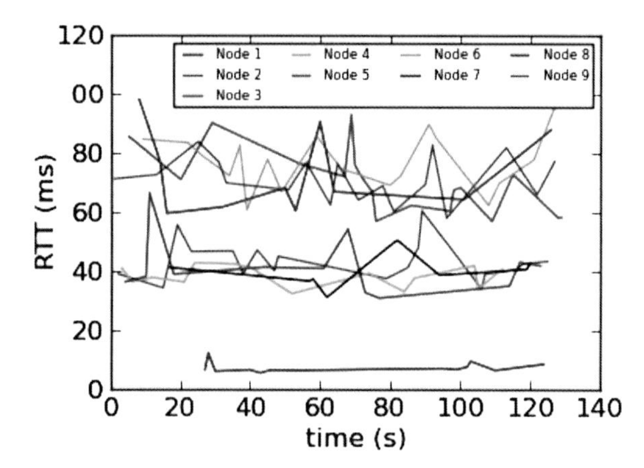

Fig. 6 Ping RTT between host machine and simulated node

4 Conclusion

This chapter describes and shows an example use of MakeSense, a framework for automating series of simulation in order to make them reproducible and reusable. The current implementation is tailored for the Cooja simualtor, and we are working on its adaptation to real testbeds. The MakeSense Python implementation for Cooja is available at: http://github.com/sieben/makesense.

This work takes place in the context of the ANR IRIS project and was partially carried out at the LINCS laboratory.

References

1. Aruliah, D.A., Brown, C.T., Hong, N.P.C., Davis, M., Guy, R.T., Haddock, S.H.D., Huff, K., Mitchell, I., Plumbley, M., Waugh, B., White, E.P., Wilson, G., Wilson, P.: Best practices for scientific computing. Comput. Res. Reposit. abs/1210.0530 (2012)
2. Eriksson, J., Österlind, F., Finne, N., Tsiftes, N., Dunkels, A., Voigt, T., Sauter, R., Marrón, P.J., COOJA/MSPSim: interoperability testing for wireless sensor networks. In: 2nd International Conference on Simulation Tools and Techniques (SIMUTools 2009), p. 27. Rome, Italy (2009)
3. Lacage, M., Ferrari, M., Hansen, M., Turletti, T., Dabbous, W.: NEPI: using independent simulators, emulators, and testbeds for easy experimentation. ACM SIGOPS Oper. Syst. Rev. **43**(4), 60–65 (2010)

Demo Abstract: Cross Layer Design for Low Power, Low Delay, High Reliability Radio Duty-Cycled Multi-hop WSNs

Eoin O'Connell and Brendan O'Flynn

Abstract We present a cross-layer approach for delivering low delay, low power and highly reliable sensor data transfer in radio duty-cycled multi-hop wireless sensor networks. We develop a novel routing metric that uses a number of weighted physical and logical parameters to enable better parent selection for routing data towards the sink. Trading latency against reliability, we demonstrate competitiveness with the state of the art, namely ORW and CTP, through direct comparison using real traces obtained from a 52 node deployment. We demonstrate reliability in excess of 99.9 % in all cases, improved energy efficiency over the state of the art, with minimal trade-offs in terms of end to end latency.

1 Introduction

In duty-cycled multi-hop WSN deployments, end-end latency is defined as the time taken for a packet generated by a node to traverse the network before it arrives at the network sink. Duty-cycled WSN deployments can introduce large end-end latencies, due to the relatively long sleep periods used in their duty-cycled nature. This work studies the effects which latency reduction protocols have on performance metrics, in an attempt to make duty cycled low power WSN's more suitable for a wider range of potentially time-sensitive and reliability focused applications.

Many potential applications for WSN's are sensitive to end-end latencies. Some specific examples of these time sensitive applications would be intruder detection systems for building security and smoke/CO detection systems. For both of the aforementioned examples, it is critical that the network sink is notified of any events within an application specific time bound. The maximum delay between detection of an event and reception at the network sink must include the per-hop latency incurred

E. O'Connell (✉) · B. O'Flynn
Tyndall National Institute, University College Cork, Cork, Ireland
e-mail: eoin.oconnell@tyndall.ie

K. Langendoen et al. (eds.), *Real-World Wireless Sensor Networks*,
Lecture Notes in Electrical Engineering 281, DOI: 10.1007/978-3-319-03071-5_8,

by each hop in the network. The most common approach to solve such an issue is to deploy a network which performs receive checks more frequently. The impact of this is increased power-consumption due to frequent receive checks.

To overcome the issues discussed above, a cross layer Media Access Control (MAC) and routing approach was designed which aims to reduce end-end latency and increase reliability. Previous latency reduction techniques choose parents solely based on the one-hop latency [3]. This can result in potentially lossy links. We present a detailed study which compares 2 versions of our work. A comparison between a protocol which optimizes for latency and one which optimizes for reliability is presented. Results are compared in terms of average end-end latencies, power consumption and most importantly reliability. The resultant protocol uses smart techniques to enable a latency aware parent selection process and enables nodes to choose parents not only based on link qualities, but also on the end-end latency which they provide.

The contributions of this work are:

- A latency aware protocol which considers end-end not just one-hop latency
- A routing mechanism/novel parent selection process which provides >99.9 % reliability
- An extremely low radio duty cycle implementation of 0.2 %
- A trade-off comparison between a latency reduction routing protocol and a reliability based protocol.

2 Design

2.1 Mac Design/Latency Awareness

Other work on latency reduction in low-power WSNs use opportunistic routing techniques such as [3] and [1]. In opportunistic routing, the first node to ACK the message becomes the parent who forwards the packet. In this work, all potential one-hop parents are probed during the learn phase and information is attained about the end-end latency which one could provide.

Our MAC protocol uses an RTS/CTS (Request-to-Send/ Clear-to-Send) preamble to contact neighboring nodes. Upon reception of a correctly addressed CTS packet, the sender proceeds by sending the payload (Payload transmission is also ACKED). With each RTS/CTS cycle having a fixed length, nodes can calculate the time offset in milliseconds between their respective wakeup schedules. The time-offsets between nodes is also equal to the latency incurred when exchanging data packets.

2.2 Parent Selection

Parent selection is based on multiple parameters. We consider multiple, multi-hop weighted parameters in selecting parents; RSSI, Hop Count, End-End Latency,

Workload, Battery Voltage, Duty Cycle and Routing Options. Nodes assign each parent an attractiveness factor and choose the best parent. These parameters are communicated in the payload ACK.

3 Deployment/ Evaluation

A total of 52 nodes were deployed in an old building spanning 3 storeys, 1 single sink node was deployed on the 3rd and top storey. The dimensions of the building are (L:60m, W:70m, H:20m), results have been compared against CTP [2] and ORW [3]. Two versions of the protocol (Version A and Version B) have been designed and experimentally verified and tested, each experiment was conducted for a minimum of 3 days. Version A was optimized for reliability and Version B for latency.

The testbed used for this experiment consists of custom devices, comprised of a PIC24F microcontroller and an SX1211 868 MHz radio transceiver. The hardware platform is pictured in Fig. 1.

In Fig. 2, results of the 3 day long experiment are presented. In Fig. 2a, averaged results for each node are depicted in terms of achieved duty cycle and end-end latency. Figure 2b, presents the reliability achieved by each node, and the average RSSI of their parents. The average RSSI of the chosen parent(s) reflects the reliability of the protocol.

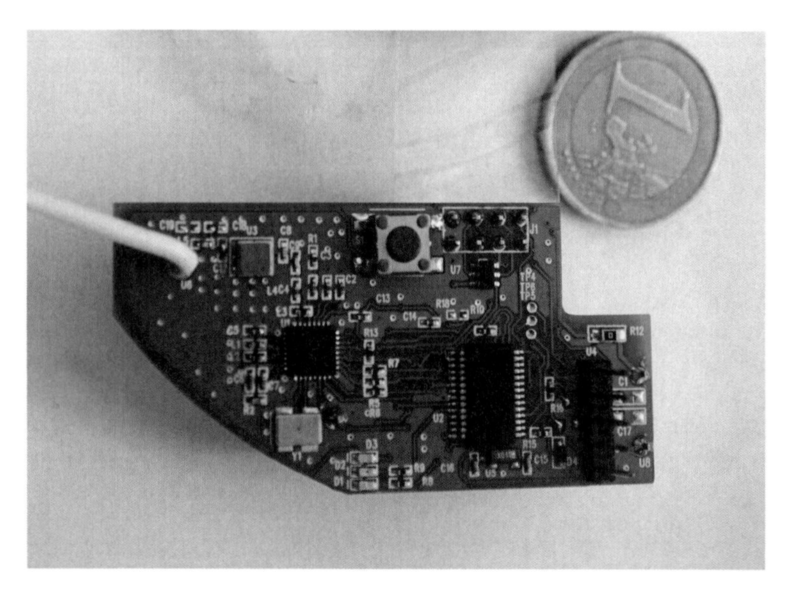

Fig. 1 Testbed WSN node, SX1211 868 MHz radio with PIC24F microcontroller. TX power: +10 dBm(@25 mA); RX current: 3 mA(@−98 dBm); sleep current:1 μA

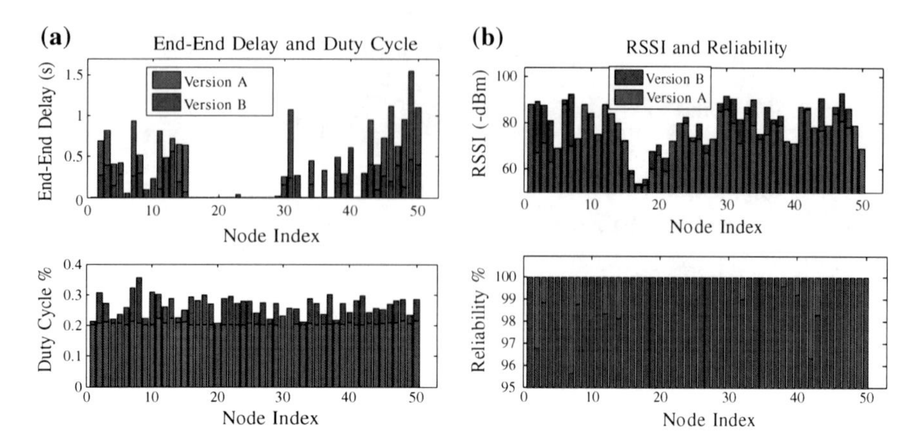

Fig. 2 Results from deployment, data for all 51 nodes plotted. **a** Latency and duty cycle. **b** RSSI and reliability

3.1 Summary of Deployment Results

Version A of the protocol achieves an average reliability of 99.983 %, Version B of the protocol achieves an average reliability a fraction lower at 99.58 %. The drop in reliability seen in Version B can be explained by Fig. 2b, this graph shows how the average RSSI of chosen parent nodes decreases. Choosing parents with lower latency but worse RSSI results in lossier links. Version A of the protocol prefers stable links over low-latency links.

Turning to the duty cycle, Version A achieves a duty-cycle of 0.207 % which is 22 % lower than that of Version B. This reduction in duty-cycle can be explained by the more stable links provided by Version A which optimizes for reliability. Compared to the state of the art, this work outperforms ORW by a factor of 4 and 3.07.

Version A also provides a very uniform distribution of the duty-cycle of all nodes under test, this would provide a more uniform lifetime for nodes in a deployment. The results presented in [3], show ORW and CTP to exhibit larger variations between the maximum and minimum duty cycles achieved by the nodes under test.

The improved latency performance of Version B seems to be an attractive option considering that there is only a slight impact on the radio duty-cycle and reliability. The 47 % reduction in average end-end latencies comes at a cost of only 0.4 % in terms of reliability and a fractional increase of 0.058 % on the overall duty-cycle figure. ORW offers on average 60 ms lower latency compared to this work, but the authors do not provide detailed reliability data.

4 Demonstration

Using a small 10–20 node network on a bench with custom nodes, we create a multi-hop network by allowing only certain nodes to communicate with the network sink. Load balancing, an energy efficient MAC and ultra reliable routing will be demonstrated. Packet latencies will be displayed as each packet arrives at the network sink, we will be able to demonstrate the latency performance of both versions of our designed protocol. By actively adding or removing nodes from the network, we can demonstrate how our load balancing algorithm works and how nodes repair damaged paths. We require only bench space and mains 230 V power sockets.

References

1. Autenrieth, M., Frey, H.: Padermac: a low-power, low-latency mac layer with opportunistic forwarding support for wireless sensor networks. In: Proceedings of the 10th International Conference on Ad-hoc, Mobile, and Wireless Networks, pp. 117–130. Springer, Paderborn, Germany (2011)
2. Gnawali, O., Fonseca, R., Jamieson, K., Moss, D., Levis, P.: Collection tree protocol. In: Proceedings of the 7th ACM Conference on Embedded Networked Sensor Systems (SenSys'09), pp. 1–14. ACM, New York (2009)
3. Landsiedel, O., Ghadimi, E., Duquennoy, S., Johansson, M.: Low power, low delay: opportunistic routing meets duty cycling. In: Proceedings of the 11th international conference on Information Processing in Sensor Networks, pp. 185–196. Beijing, China (2012)

Poster Abstract: Outdoors Range Measurements with Zolertia Z1 Motes and Contiki

Marie-Paule Uwase, Nguyen Thanh Long, Jacques Tiberghien, Kris Steenhaut and Jean-Michel Dricot

Abstract Practically useful outdoor transmission ranges have been determined experimentally for Zolertia Z1 motes running Contiki. Both internal and external antennas were tested. The analog and digital quality of the received signal as well as the packet delivery rate have been measured. The influence of transmission power, bushes and weather conditions have been explored. Internal antennas were excessively sensitive to orientation, while external antennas allowed reliable and rather stable communications over distances between motes of up to 150 m. Radio duty cycling was responsible for some, still partially unexplained, packet losses.

1 Motivation

To study experimentally the interactions of the Radio Duty Cycling (RDC), MAC and Routing over Low power and Lossy Networks (RPL) protocols for Wireless Sensor and Actuator Networks (WSAN), a test facility is being built in the garden of a countryside house because observations will be easier to interpret far from potentially disturbing radio transmitters and from reflecting structures causing multipath interferences. The facility consists in a WSAN, with a dozen static motes, positioned all over a 1,000 m^2 lawn, one mobile mote, affixed on a robotic lawn mower, and one gateway mote, located near a veranda, where computers and test equipment are housed. The position and the transmission power of the static motes should force the RPL routing algorithm to select as many as possible multi-hop

M.-P. Uwase (✉) · N. T. Long · J. Tiberghien · K. Steenhaut
Vrije Universiteit Brussel, ETRO, Pleinlaan 2, 1050 Brussels, Belgium
e-mail: marie.paule.uwase@vub.ac.be

M.-P. Uwase · J. Tiberghien · J.-M. Dricot
Université Libre de Bruxelles, OPERA, Av. Franklin Roosevelt 52, 1050 Bruxelles, Belgium

M.-P. Uwase
National University of Rwanda, Butare, Rwanda

K. Langendoen et al. (eds.), *Real-World Wireless Sensor Networks*,
Lecture Notes in Electrical Engineering 281, DOI: 10.1007/978-3-319-03071-5_9,

routes. As a survey of the relevant literature [1–6] did not provide sufficient practical data to implement this last requirement, we undertook ourselves a practical study of radio links, inspired by the experimental work reported in [5] and [6], but using modern motes available on today's market. We believe some of our observations can be useful for anybody designing a WSAN.

2 Experimental Set-Up

Most of the experiments with WSANs in our environment are done with Zolertia Z1 motes [8] running Contiki [9]. The Z1 exists in two versions: one with a built-in antenna, another with an external antenna. Both were investigated. The quality of a radio link can be characterized by three indicators. Two of them are evaluated by the radio receiver itself: the analog Radio Signal Strength Indicator (RSSI) and the digital Link Quality Indicator (LQI) [7] estimated by the demodulator. The third is directly relevant to the applications: the Ratio of correctly Delivered Packets (PDR). Studies showing that these three indicators do not correlate very well can be found in [8] and [9]. Therefore, whenever relevant, the values of the three will be given. To evaluate these indicators, a point to point unicast link was set up, using the communication functions provided by Contiki rime. The motes were mounted on 1m wooden poles by means of nylon straps. The battery powered transmitter could freely be positioned anywhere in the garden and given any orientation. The receiver was positioned at 1m from the veranda and was connected by a USB cable to a PC. All range measurements were done along three straight lines, with no obstacles between sender and receiver. For avoiding interference with possible WiFi networks all tests were performed in channel 26, the channel which is least susceptible to WiFi interference. The absence of interfering radio fields was monitored during the experiments by means of a SPECTRAN HF-4060 portable spectrum analyzer from Aaronia. Radio traffic was monitored by a TI packet Sniffer with a cc2531DK dongle.

3 The Measurements

To test a link, the sender transmitted every second a 45 bytes packet. For 50 consecutively received packets the values of RSSI, LQI and the sequence number of the packet were recorded by the PC connected to the receiver. The global values of RSSI and LQI were obtained by averaging the recorded values, while the PDR value was computed by dividing the number of received packets by the number of sent packets. The latter was obtained from the sequence number included by the sender in each packet. For these tests, the radio was continuously left on.

3.1 Directivity Measurements

The first test aimed at comparing the motes with built-in antenna and those with an external vertical $\lambda/4$ whip antenna. Preliminary tests showed that for links exceeding 1 m, the motes with built-in antenna needed to be held vertically. For evaluating the horizontal directivity, the properties of the radio link were measured with 8 different orientations of the sender with respect to the receiver. This clearly showed that the internal antenna of the Z1 motes is far from omnidirectional (20 dB difference between the best and worst direction) and that transmission in front of the mote is especially poor. It was suspected that the presence of batteries in front of the internal antenna was the cause of this poor performance, but removing the batteries proved this hypothesis wrong. As the internal antenna would be a source of unwanted and hard to explain artifacts, it was decided to use exclusively motes with an external vertical antenna. These antennas appeared to be horizontally omnidirectional and to have a main vertical radiation pattern covering approximately 20°.

3.2 RSSI Calibration

The next test consisted in verifying the accuracy of the RSSI indications by the receivers. One by one, all motes, under identical conditions, received 50 packets from a single sender. The average values of RSSI for each mote were compared. They all were within a 2 dB margin, except for one mote that gave a 20 dB lower value. The 2 dB margin complies with the CC2420 specifications [7] but the 20 dB were considered an indication of malfunctioning. The antenna or its cable was suspected but substitution proved the problem was elsewhere.

3.3 Range Measurements

First, the maximal range was evaluated at 0 dBm in an open field. It was found to be 160 m. Thereafter the quality of radio links was measured several times in three different directions, over distances ranging between 3 and 21 m with steps of 3 m and with four different levels for the transmitted power (0, −10, −15 and −23 dBm). An almost monotonous, but far from predictable, decrease of the RSSI values in function of the distance could be observed, but large differences between measurements done in different directions and/or weather conditions confirmed the affirmation that RSSI values cannot be used to estimate distances [10]. The LQI values did not show a similar pattern: often they were better at medium distances (12 m) than nearby (3 m). However, most observed values were above 90 which corresponds to a good digital link. The quality of these digital links was confirmed by the PDR values, which were 100 % for all links, except for one 21 m link (with −25 dBm sending power) where RSSI (−93 dBm), LQI (60) and PDR (<10 %) clearly showed that this was out of range.

3.4 The Influence of Radio Duty Cycling

When range measurements were started in 2012, naively the default configurations of the motes and their Contiki operating system (power = 0 dBm, MAC = CSMA, RDC = contikimac, 8 Hz [11]) were used. Despite good RSSI and LQI figures, abnormally low PDR values were obtained. Disabling the power saving radio duty cycling (RDC) by replacing the contikimac driver by the nullrdc driver brought the PDR to 100 % in almost all experiments, as reported above. The very low RDC was found to be due to a timing bug in the Z1 version of the contikimac driver. The bug was reported in the appropriate forum [12] and corrected. This resulted in a PDR close to 100 % for RSSI levels better than −75 dBm, but degrading rapidly beyond that limit (with a RSSI of −80 dBm, PDR is as low as 16 %). This degradation is probably due to the Clear Channel Assessments threshold set at −80 dBm used to detect arriving packets.

The degradation due to power savings was explored further by adding to the range measurements described above, tests with the three different RDC protocols readily implemented in Contiki (contikimac, Xmac [13] and Low Power Probing (LPP) [14]). With Xmac, no degradation of the PDR could be observed, even with an RSSI as low as −84 dBm. The case of LPP is quite different: below −50 dBm, messages are sometimes duplicated or triplicated because the LPP driver in the sender ignores some acknowledgments received and forces the CSMA MAC layer to retry or to drop the packet. This is clearly due to a timing issue, because the problem disappears when the DEBUG option is enabled in the LPP driver.

4 Conclusions

The Zolertia Z1 motes can ensure reliable radio communications over links with attenuations that do not exceed 90 dB. In an open dry grass field, this corresponds to a distance of some 180 m when the vertical whip antennas provided with the motes are used. Wet grass or moderate rain can cost up to 15 dB and a single leafy bush on the path costs 15 dB, meaning that radio links exceeding 150 m should be avoided. The Z1 mote is also available with an internal antenna. Compared with the whip antenna, this costs between 25 and 45 dB, depending on the relative orientation of the motes, practically restricting the motes with an internal antenna to desktop experiments. The above measurements were conducted with the radios always on, and resulted in PDRs very close to 100 %. Enabling energy saving RDC protocols changed this figure, especially over long links, but we suspect this is due to some inadequate tunings of parameters built into the Contiki implementation of these protocols.

References

1. Trenkamp, P., Becker, M., Goerg, C.: Wireless sensor network platforms datasheets versus measurements. In: Annual IEEE Conference on Local, Computer Networks, pp. 966–973 (2011)
2. Srinivasan, K., Ditta, P., Tavakoli and P. Levis: An empirical study of low power wireless. ACM Trans. Sens. Netw. **6**(2) (2010)
3. Weilian, Su: Mohamad Alzaghal: channel propagation characteristics of wireless MICAz sensor nodes. Ad Hoc Netw. **7**, 1183–1193 (2009)
4. Verdone, R., Gezer, C., Flavio, F.: The impact of realistic footprint shapes on the connectivity of wireless sensor networks. In: Proceedings of WSNPERF 2009, Pisa. ACM, Italy (2009)
5. Anastasi, G., Conti, M., Falchi, A., Gregori, E., Passarella, A.: Performance measurements of mote sensor networks, Department of Information Engineering, University of Pisa and CNR-IIT Institute. http://citeseerx.ist.psu.edu/viewdoc/download?doi=10.1.1.74.8076&rep=rep1&type=pdf (2004). Accessed 24 July 2012
6. Niels, R., Gertjan, H., Koen, L.: Link layer measurements in sensor networks. Faculty of Electrical Engineering, Mathematics and Computer Science, Delft University of Technology, The Netherlands. http://www.st.ewi.tudelft.nl/ koen/papers/mass04.pdf (2004). Accessed 24 July 2012
7. CC2420 Datasheet, Chipcon instruments from Texas Instruments. http://www.ti.com/lit/ds/symlink/cc2420.pdf (2012). Accessed on 24 July 2012
8. Holland, M.M., Aures, R.G., Heinzelman, W.B.: Experimental investigation of radio performance in wireless sensor networks. Department of Electrical and Computer Engineering, University of Rochester, Rochester New York. http://www.ece.rochester.edu/projects/wcng/papers/conference/holland_secon06.pdf (2010). Accessed 24 July 2012
9. Liu, T., Cerpa, A.-E.: Foresee (4C): Wireless Link Prediction Using Link Features. Electrical Engineering and Computer Science, University of California–Merced. http://andes.ucmerced.edu/papers/Liu11a.pdf (2012). Accessed 24 July 2012
10. Z1 Datasheet, Zolertia. http://www.zolertia.com/ (2010). Accessed March 2010
11. Dunkels, A.: The ContikiMAC radio duty cycling protocol. Technical Report T2011:13, Swedish Institute of Computer, Science December (2011)
12. https://github.com/contiki-os/contiki/pull/224
13. Michael, B., Gary, V.Y., Eric, A., Richard, H.: X-MAC: A Short Preamble MAC Protocol for Duty-Cycled Wireless Sensor Networks. Department of Computer Science University of Colorado, USA. http://www.cse.wustl.edu/lu/cse521s/Papers/x-mac.pdf
14. R. Musaloiu-Elefteri, C. Liang, A. Terzis. Koala: Low power probing—ultra-low power data retrieval in wireless sensor networks. IPSN 2008

Poster Abstract: iBAST—Instantaneous Bridge Assessment Based on Sensor Network Technology

Richard Mietz, Carsten Buschmann, Dennis Boldt, Kay Römer and Stefan Fischer

Abstract This poster presents the iBAST project, the goal of which is the continuous monitoring of highway bridges with sensor network technology. Currently working on a first individual solution, the long term goal of the project, though, is to move from individual case studies towards a comprehensive monitoring of bridges, by developing a monitoring system which can be easily and cost-effectively customized for every bridge. This poster abstract presents the system architecture and implementation as well as first results.

1 Introduction

Bridges are subject to a continuous process of aging and damage and require continuous monitoring and maintenance, which is associated with considerable costs. Under acute pressure on costs, these measures are often minimized, which in the past led to the expansion of existing damage—and even worse, to emergency measures such as suspension of lanes, closedowns for trucks and buses, or even complete closedowns of the whole bridge.

We believe that such situations can be prevented by using modern wireless sensor network (WSN) technology. The aim of this project is therefore to design a cost

This work is partially funded by the Federal Highway Research Institute of the Federal Republic of Germany (Förderkennzeichen 88.0122/2012, iBAST).

R. Mietz (✉) · D. Boldt · S. Fischer
Institute of Telematics, University of Lübeck, 23538 Lübeck, Germany
e-mail: mietz@itm.uni-luebeck.de

C. Buschmann
coalesenses GmbH, 23562 Lübeck, Germany

K. Römer
Institute for Technical Informatics, Graz University of Technology, 8010 Graz, Austria

K. Langendoen et al. (eds.), *Real-World Wireless Sensor Networks*,
Lecture Notes in Electrical Engineering 281, DOI: 10.1007/978-3-319-03071-5_10,

effective bridge management and maintenance through continuous monitoring with advanced information technology. Actually, even more ambitious, the long term goal of the project is to move from individual case studies such as [1–3] towards a comprehensive monitoring of bridges using WSNs. This requires the reduction of effort of using a WSN for monitoring a bridge—which is now measured in person-years of engineers and scientists. Because each bridge is unique with individual damage and peculiarities, the goal of the project is to develop a monitoring system which can be easily and cost-effectively customized for every bridge. This is achieved by building a modular system consisting of a set of ready-made hardware and software components which can be assembled in the way needed for the specific bridge. The functioning of the selected components is easily adapted to the monitoring task of the bridge. A central objective of the research and development in this project is therefore a design which can be installed and configured by suitable trained engineers.

2 Architecture and Implementation

In order to implement this approach, we designed a layered architecture which is shown in Fig. 1. It consists of the WSN application running on the developed hardware, the backend with database, and the user interface (UI):

Hardware The *hardware platform* was designed following a modular approach (see Fig. 2). The main board on the left contains the basic functionality, such as the micro controller and radio, energy management, peripherals such as external storage (flash and/or SD card), a low power accelerometer, a USB interface and an optional UI consisting of a display and a rotary/push controller. The device (see Fig. 3, depicted with attached external pressure sensor) can be powered from internal batteries (up to 136 Wh) or external supplies (e.g., a solar panel or 3–24V DC), and can provide configurable voltages to external components. To adapt to different measurement tasks and sensor types, the main board can be extended by two expansion boards (cf. Fig. 2 on the right) that can carry internal sensors (e.g., a high precision inclinometer) or interfaces to external sensors (e.g., high-resolution ADC for strain gauges) and communication devices (e.g., GPRS/UTMS modem). This allows to use completely different sensor types on each of the expansion slots, i.e., multi-sensor nodes can

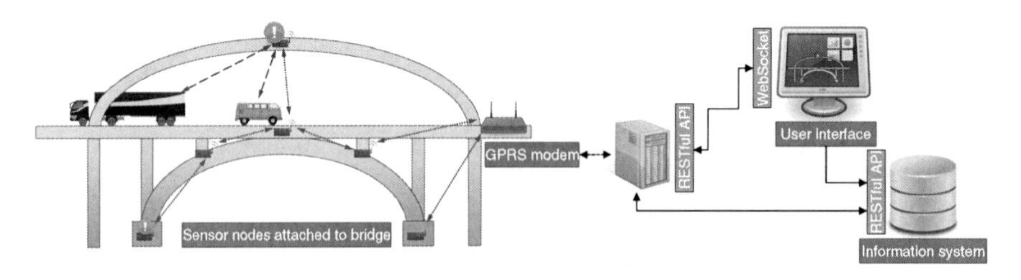

Fig. 1 Architecture of the iBAST-system

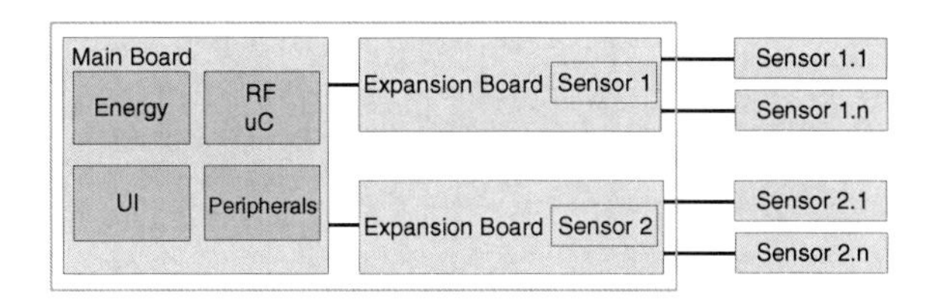

Fig. 2 Hardware architecture

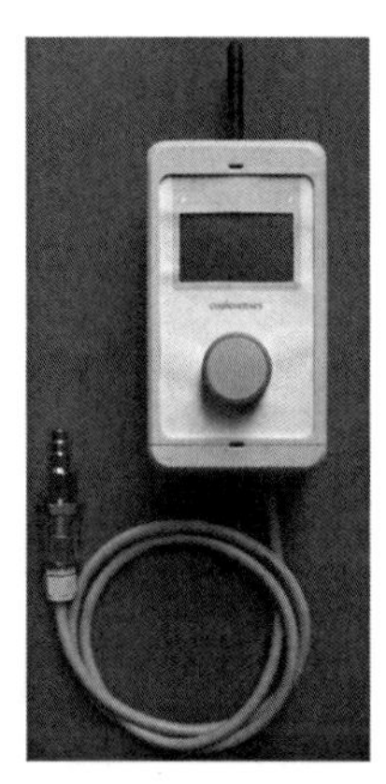

Fig. 3 Sensor node

be created. Within the project iBAST, we will sense impact factors (e.g., temperature, humidity, wind, traffic) on the one hand and construction responses (e.g., tilt, displacement, crack opening/closure) on the other in order to be able to correlate them.

WSN Application The *WSN application* interprets scripts defined by users. The scripting language allows defining events (e.g., car passes the bridge, sensor measure exceeds threshold, or a specific time), behaviors (e.g., sense, log, and send values), and scopes (a set of sensors defined by addresses or sensor type). Distributed if-then rule evaluation of logical formulas over events trigger behaviors for the sensors of a given scope. Additionally, a compression script (based on [4]) allows lowering the data size by reducing the level of detail of sensed data. Furthermore, an accuracy-driven time synchronization protocol allows adjusting the accuracy of the nodes' clocks to the needs of the deployed sensor types and the current script interpreted by the application to reduce resource consumption.

Backend The *backend* handles the communication between WSN and UI by translating between protocols and (de-)serialization of verbose data formats (used on the Internet) and binary representations (used in the WSN). Moreover, it manages data of individual bridges, i.e., static metadata, real-time sensor data, and data provided by users (e.g., reports and photos). Data collected by the sensors is interpreted based on stored bridge damage models, and a grade is computed, in compliance with the DIN standard of [5], which reflects the state of the bridge. Since this is done continuously, changes in this state early thus enabling timely counter measures.

User Interface The *UI* is a platform independent single-page web application realized with HTML5 technologies. Hence, observing the live state of a bridge is possible without polling the backend continuously. It is specifically designed for the workflow of and easy use by the target audience of structural engineers. Users can view information of maintained bridges as well as add arbitrary information like notes or files such as pictures, blueprints or reports. Visualizations of current and past sensor values (e.g., as charts) allows inspecting and tracking of the bridge state for any given time interval. Furthermore, a rights management system can provide different views on the data for the maintainer, a subcontractor, or the public. Finally, new scripts can be created and uploaded to the WSN to change or adapt monitoring goals.

3 Case Study

The state of Schleswig-Holstein is providing a prestressed concrete road bridge near Lübeck which spans a federal highway to conduct a case study to test the developed hardware and software components under realistic weather, traffic, and damage conditions.

We already carried out successful gluing tests to evaluate the influence of different weather conditions on the durability of the mounting of sensor nodes to the bridge. A second experimental series we undertook has given insight into radio propagation characteristics at the bridge.

Next, the deployment of a sensor node with different environmental sensors and the GPRS modem will allow testing of the first versions of the WSN application, backend, and UI. These components will then be incrementally enhanced and evaluated. Furthermore, we will deploy additional sensor nodes step-by-step to measure and analyze, among others, interactions between different environmental parameters, traffic, and crack growth. Finally, our goal is a 1 year full deployment of the bridge using approximately 20 sensors nodes.

4 Conclusion

We presented the iBAST project in which we develop an easily and cost-effectively customizable monitoring system for bridges with the help of WSN technology which helps bridge maintainers to detect critical damages earlier and easier.

In the end of the project we expect to have all components fully tested and are able to provide building blocks for arbitrary concrete bridges. The components should then be able to self-sustainingly measure, aggregate, and correlate data of different sensors to compute a grade reflecting the structural health of the bridge for a period of 6 years which is the typical inspection interval of bridges in Germany.

References

1. Chebrolu, K., Raman, B., Mishra, N., Valiveti, P.K., Kumar, R.: Brimon: a sensor network system for railway bridge monitoring. In: Proceedings of the 6th MobiSys (2008)
2. Feltrin, G., Meyer, J., Bischoff, R., Motavalli, M.: Long-term monitoring of cable stays with a wireless sensor network. Struct. Infrastruct. Eng. **6**(5), 535–548 (2010)
3. Kim, S., Pakzad, S., Culler, D., Demmel, J., Fenves, G., Glaser, S., Turon, M.: Health monitoring of civil infrastructures using wireless sensor networks. In: Proceedings of the 6th IPSN (2007)
4. Mietz, R., Römer, K.: Work in progress: resource-aware fault localization in large sensor networks. In: Proceedings of the 8th DCOSS (2012)
5. Deutschland. Bundesministerium für Verkehr, B.u.W.: Richtlinie zur einheitlichen Erfassung, Bewertung, Aufzeichnung und Auswertung von Ergebnissen der Bauwerksprüfungen nach DIN 1076: RI-EBW-PRÜF; Verkehrsblatt-Dokument Nr. B 5236. Verkehrsbl.-Verlag (1998)

Demo Abstract: SmartSync; When Toys Meet Wireless Sensor Networks

Fiona Edwards-Murphy, Michele Magno, Aidan Frost, Amy Long, Naomi Corbett and Emanuel Popovici

Abstract Since ancient times, toys have captured the imagination of children the world over with major roles in learning and entertainment. This chapter presents SmartSync, a platform for interactive toys using ultra low power Wireless Sensor/Actuation Network (WSN) technologies. These technologies are integrated into a number of off-the shelf micro-robots and colour LED arrays, making them fully autonomous. A demo system has been developed in which wireless micro-robots and the colour LED arrays interact with a wireless microphone to create an interactive concert scenario. The LED lights and dancing robots are synchronised with music through the WSN. In the final demonstration, the effectiveness of the dancing robots and lights are shown when they respond to traditional Irish music. The WSN interface added improves versatility and accessibility to such toys at a very low cost and low power overhead. This work represents a first step in the development of intelligent interfaces for enhanced interaction with toys for children, especially for those with disabilities.

F. Edwards-Murphy (✉) · M. Magno · A. Frost · A. Long · N. Corbett · E. Popovici
Department of Electrical and Electronic Engineering, University College Cork, Cork, Ireland
e-mail: f.edwardsmurphy@umail.ucc.ie

M. Magno
e-mail: m.magno@ucc.ie

A. Frost
e-mail: a.frost@umail.ucc.ie

A. Long
e-mail: a.d.long@umail.ucc.ie

N. Corbett
e-mail: naomi.corbett@umail.ucc.ie

E. Popovici
e-mail: e.popovici@ucc.ie

K. Langendoen et al. (eds.), *Real-World Wireless Sensor Networks*,
Lecture Notes in Electrical Engineering 281, DOI: 10.1007/978-3-319-03071-5_11,

1 Introduction

Recent technological advances have led to the development of embedded sensing, computing and communication devices; these are becoming an essential part of daily life. These ubiquitous platforms, known as Wireless Sensor Networks (WSN) have been recognised as a fundamental technology for a large variety of applications including smart homes, personalised healthcare, security etc. [1]. WSN technology has gained popularity in recent times, as it supports a wide range of applications.

Toys play an important role in our lives. From a young age we enjoy the company of toys and use them for entertainment and learning. This project aims to expand the accessibility and interaction experience with toys using the latest emerging wireless sensors, sensor interfaces and low power networking protocols. Heterogeneous WSN allow novel intelligent interfaces which are non-intrusive, natural, and empowering for children and could potentially have a large positive impact on everyday life [2].

The aim of this work is to design a scalable, low power infrastructure for collaborative toys using WSN technologies. A system view is taken, combining ultra-low power processing and communication features of WSNs, these are integrated onto three low cost versatile micro-robot toys [3]. The platform used to develop all nodes is Texas Instruments' (TI) EZ430-RF2500 [4] which includes a 2.4 GHz RF transceiver, MSP430F2274 microcontroller and power management unit; this allows network implementation using the SimpliciTI compliant protocol stack.

Three robotic spiders are modified, allowing remote WSN control and interoperability. These toys are deployed on a stage in a concert scenario with a custom built low power, low cost wireless sound detection and light system. The system is capable of providing physical and visual feedback to music. A similar approach could be used to respond to gestures making it ideal for education and a wide range of therapies [5].

1.1 System Integration

In order to provide a demo, a stage was designed. The floor of the stage was built from a thin sheet of aluminium, providing reflections from the LED lights as well as a drum sound effect which is a key aspect of Irish dancing. Three wireless LED arrays were mounted at the top of the stage on metal railings. The three modified wireless Hexbug toys were placed on the stage where they dance. Finally, the wireless coordinator node with microphone sensor was placed in the vicinity of a source of music.

2 Implementation

In this chapter a heterogeneous wireless network is presented with three types of WSN nodes: a motor driver node, an audio sensing node and an LED array node.

2.1 Wireless Micro-Robot Node

Hexbugs are a range of small, clever robotic toys designed and manufactured by Innovation First [6]. The Hexbug "Spider" was selected as its movement can be controlled by two DC motors. This platform is adapted for remote wireless control by replacing its original motor control with a custom motor control WSN node.

The architecture of robot node (Fig. 2a) includes an EZ430-RF2500 and a motor interface. The motors are driven using a dual H-bridge TB6612FNG. This board controls two DC motors at a constant current of up to 1.2 A. The motor speed is controlled using a PWM signal generated by the microcontroller's on board timers. Low power, accurate wireless motor control is a key feature of robotic toys as it allows for toys to provide immediate feedback to stimulus. The developed firmware implemented several functions to make the spider dance in different styles in response to network commands and to create a visually stimulating and autonomous reaction to the music.

During assembly, the unit was made as compact as possible to prevent the Spider loosing balance. A small, light, rechargeable lithium polymer battery was used to power the robot. The lithium battery will also allow a miniaturised charging circuit, which will decrease the long term cost of the system.

2.2 Wireless Microphone Circuit

A music detection unit was designed and developed to allow the toys and LED's to react to music. Figure 2b shows how the music sensor is interfaced with the EZ430-RF2500. The wireless microphone node becomes the coordinator of the star topology SimpliciTI network (Fig. 1).

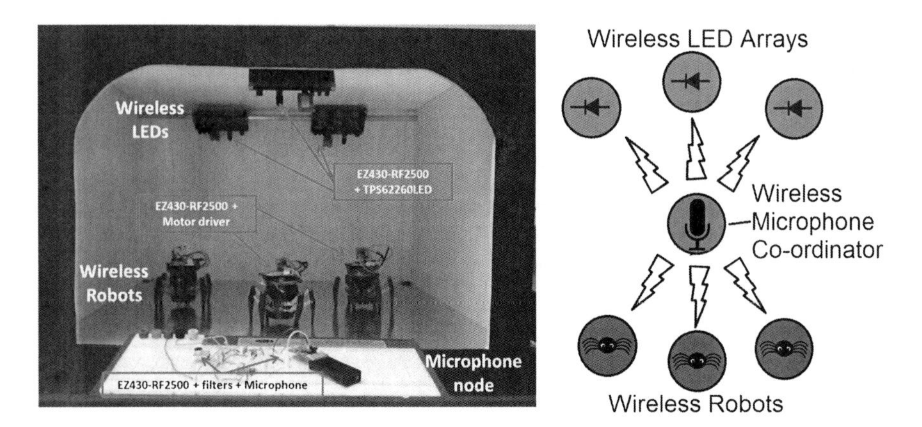

Fig. 1 Final smartsync system deployment and system architecture

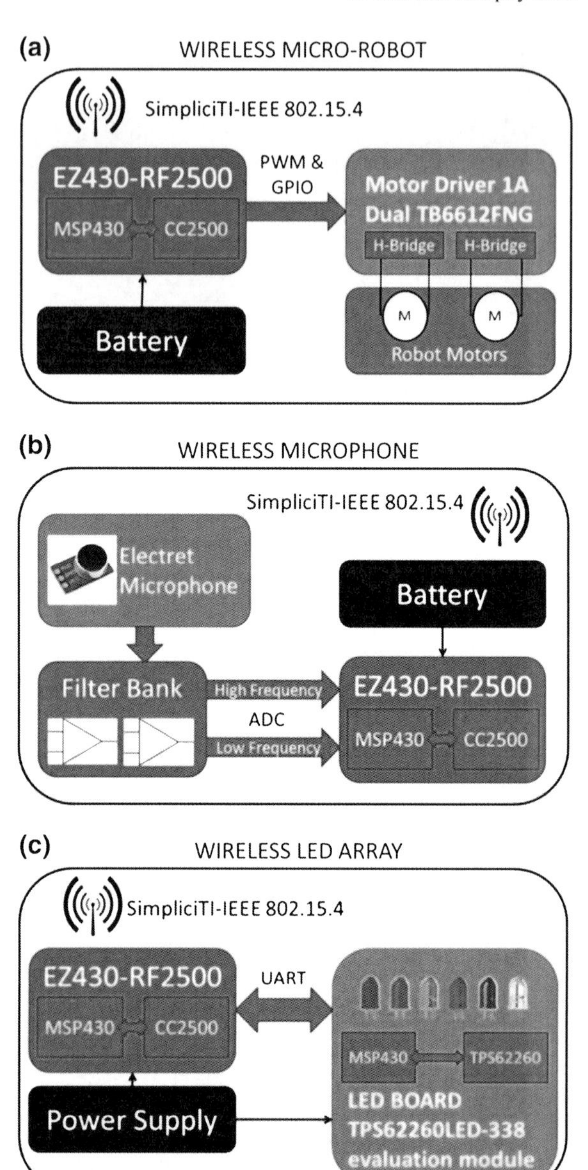

Fig. 2 Architecture of: **a** Micro-robot node, **b** Microphone node, **c** LED node

Graphic equalisers use banks of filters to divide music into frequency bands for individual processing. The bands are often displayed in a graph using LED or LCD displays [7]. This method is used in the SmartSync sensor to separate detected music, these bands are analysed by the WSN node, which controls the LEDs and robots. In the circuit the amplified output of an electret microphone is taken. This output is fed into two filters, low-pass and high-pass; this separates the "bass" and "treble" notes.

The sensor network response becomes more engaging and attractive as it responds differently to varying instruments and styles of music.

The WSN node in this unit was using the EZ430-RF2500, both audio bands provided by the audio sensor are sampled by the on board 10-bit ADC. The amplitude of each input is compared against a series of thresholds. Based on the results the RF transceiver disseminates a related command throughout the network.

2.3 *Wireless LED Arrays*

The concert scenario would not be complete without providing proper lighting solutions. This work considers an array of RGB LEDs, wirelessly controlled with a WSN node to provide a disco type effect. The lights are synchronised with the movements of the micro-robots through the wireless interface.

In order to create this node, the Texas Instruments TPS62260LED module and EZ430-RF2500 were used (Fig. 2c). The TPS62260LED drives its three on board LED's using TPS62260 2.25 MHz 600 mA step down voltage converters, which are controlled by an MSP430F2131 microcontroller. The intensity of the LED's can be varied to produce a variety of coloured transitioning light effects. The EZ430-RF2500 provides networking capabilities. The two MSP430's are interfaced using UART. The MSP430F2131 drives the buck converters in response to commands on the network.

3 Conclusions

A versatile, low cost and low power platform for collaborative toys using WSN technologies has been developed. The demonstration shows the feasibility and fundamental features of integrating WSN technology with low cost, battery operated toys in a classical example of a cyber-physical system. It is demonstrated that WSN is an enabling technology for real-world collaborative toys. Further work will expand the system's range of sensory abilities, collaborative signal processing abilities and communications; as well as providing further power optimisation [8].

Acknowledgments The authors thank Michael O'Shea and Tim Power in the Dept. mechanical workshop, and TI University Programme for their support.

References

1. Shiwei, Z., Haitao, Z.: A review of wireless sensor networks and its applications. In: 2012 IEEE International Conference on Automation and Logistics (ICAL), pp. 386–389 (2012)
2. Schwarz, J.: A grammar-based system for game playing with a sensor network. http://www.eng.yale.edu/enalab/publications/Jonathangrammars.pdf (2006)
3. Guerrero, P., Cilia M., Buchmann, A.: Relying on wireless sensor networks to enhance the RC-gaming experience. In: 3rd International Workshop on Pervasive Gaming Applications (PerGames 2006), Ireland (2006)
4. Jelicic, V., Magno, M., Brunelli, D., Bilas, V., Benini, L.: An energy efficient multimodal wireless video sensor network with eZ430-RF2500 modules. In: Proceedings of the IEEE 5th International Conference on Pervasive Computing and Applications, pp. 161–166 (2010)
5. Banerji, S.: A unified, neuro-physio platform to facilitate collaborative play in children with learning disabilities. In: IEEE International Conference on Rehabilitation Robotics, pp. 912–917 (2009)
6. Innovation First, http://www.innovationfirst.com/ (2013)
7. Solomon, C.W.: A 12 band microprocessor controlled graphic equalizer display filter IC. IEEE Trans. Consum. Electron. 663–670 (1991)
8. Popovici, E., Magno, M., Marinkovic, S.: Power management techniques for wireless sensor networks: a review. In: 2013 5th IEEE International Workshop on advances in sensors and interfaces (IWASI), pp. 194–198 (2013)

Poster Abstract: Link Quality Estimation—A Case Study for On-line Supervised Learning in Wireless Sensor Networks

Eduardo Feo-Flushing, Michal Kudelski, Jawad Nagi, Luca M. Gambardella and Gianni A. Di Caro

Abstract We focus on the implementation issues of on-line, batch supervised learning in computationally limited devices. As a case study, we consider the use of such techniques for link quality estimation. We compare three strategies for the on-line selection of the data samples to be kept in memory and used for learning. Results suggest that strategies that keep balanced the set of training samples in terms of ranges of target values provide better accuracy and faster convergence.

1 Introduction

In real-world scenarios, wireless sensor networks (WSNs) might need to operate in harsh conditions, facing dynamic variations in the environment. In many of these situations, it is useful, or even required, to model the variability and the uncertainty of the environments in order to understand the current situation and/or to make predictions about future events. To this end, *machine learning* techniques have been proven

This research has been partially funded by the Swiss National Science Foundation (SNSF) Sinergia project SWARMIX, project number CRSI22_133059.

E. Feo-Flushing (✉) · M. Kudelski · J. Nagi · L. M. Gambardella · G. A. Di Caro
Dalle Molle Institute for Artificial Intelligence (IDSIA), Manno, Switzerland
e-mail: eduardo@idsia.ch

M. Kudelski
e-mail: michal@idsia.ch

J. Nagi
e-mail: jawad@idsia.ch

L. M. Gambardella
e-mail: luca@idsia.ch

G. A. Di Caro
e-mail: gianni@idsia.ch

K. Langendoen et al. (eds.), *Real-World Wireless Sensor Networks*,
Lecture Notes in Electrical Engineering 281, DOI: 10.1007/978-3-319-03071-5_12,

to be a flexible and effective approach to construct models able to capture complex relationships among many different variables. Unfortunately, these techniques often consume considerable amounts of computational resources. This might constitute a barrier to the use of machine learning in WSNs, since these networks are commonly made of devices with limited computational capabilities, both in terms of processing and storage. These limitations raise the need to optimize, or set strict bounds to, the consumption of computational resources during implementation and execution of machine learning algorithms, effectively balancing performance and consumption of resources.

Among the large variety of machine learning techniques, *supervised learning* plays a central role, both historically and in terms of actual applications. In supervised learning a system automatically learns on the basis of a set of labeled training data given as input. In WSNs facing variable situations, on-line supervised learning schemes in which the system adaptively *re-learn* or learn *incrementally* can turn out to be very useful in practice. In a very broad sense, the implementation of on-line supervised learning techniques, can be realized using two main approaches. In the *batch learning* approach, iteratively, a node first gathers batches of data to learn from and use them to re-build (from scratch) the prediction model. For instance, Support Vector Machines can be conveniently used at this aim. In the *incremental approach*, a node incrementally updates the model after a new training sample is gathered. Although incremental approaches allow for significant savings in computing resources (mainly processing), their internal models might continually increase on size as new data samples arrive. Moreover, it might difficult to restrict the size of those models, or to control the trade-off between memory requirements and prediction accuracy. An example of incremental approaches which suffers from these problems is the popular *Locally Weighted Projection Regression* (LWPR) [1]. Instead, in batch learning, resource consumption is controllable by deciding/limiting the maximum size of the training set and the frequency of rebuilding the model, making them more suitable to be implemented in computationally limited devices such as WSNs.

In this work, we focus on the implementation issues of on-line, batch learning in computationally limited, embedded networked devices. As a case study, we consider the use of such techniques for *link quality estimates*. This is motivated by our recent work [2] in which we proposed LQL, a protocol for the *on-line supervised learning* of link quality estimates in wireless networks. LQL relies on passive channel monitoring to gather information about the quality of the current wireless links. This information is used to learn a regression mapping between the local network configuration and the expected link quality. Considering packet reception rate (PPR) as quality metric, it was shown that LQL allows to obtain fast and reliable estimates both in simulation and on a real testbed. In [2] we used LWRP for incremental learning and we did not address the issue of the limited computational capabilities, especially for *memory*, which we address here: as data is continually gathered, a node needs to decide which data to include in its training set and which not, since not all training information can be stored because of limited memory. We study different selection strategies and compare their performance using a dataset obtained from a real WSN testbed.

2 On-line Training Data Selection for Batch Strategies

As each new sample arrives, we face the decision of whether this sample should be included or not in the training set, and, in the positive case, which sample must be then removed (if needed) to keep the set size within the prescribed limits. We evaluated three simple, but representative, strategies. The first is called FIFO, and it treats the current training set as a *first in, first out* queue: a new sample is always accepted and the oldest sample is always removed. In the RANDOM strategy, the decisions are taken according to a probabilistic rule. The probability of a new sample entering the set is 0.5. In the positive case, if the set is full, an existing sample to be removed is chosen uniformly at random. The SLOTTED strategy consists of uniformly partitioning the [0, 1] interval of possible link quality values in $n = 10$ intervals, ($q_1 = [0, 0.1], \ldots, q_{10} = [0.9, 1]$). The training set is partitioned accordingly into n slots. Each slot i contains the samples whose link quality value (PRR in our case) falls inside the corresponding interval q_i. The strategy, aims to maintain a set of training data that covers more or less uniformly the range of link quality values (i.e. good and bad links should be equally represented in the set). All slots are managed with FIFO. For all strategies, the first model is built when the training set reaches the maximum allowed size. Then, after replacing half of the samples, a new model is built.

3 Experimental Evaluation

To implement batch learning, we used *Gaussian Process Regression* (GPR) [3], a nonparametric, flexible regression method characterized by good prediction accuracy, computational tractability, and its relatively simple implementation [4].

We consider a dataset obtained from the INDRIYA Testbed, a 3D WSN deployed across three floors of the School of Computing at the National University of Singapore [5]. The network is composed of 139 TelosB sensor motes. Nodes run the TinyOS operating system and are programmed in the NesC programming language. An implementation of LQL was used to collect data, which consisted in vectors of network features playing the role of measurable parameters determining the quality (PPR) of a link. Since INDRIYA nodes are static, in order to maximize the number of observed topologies and simulate changes related to mobility and/or radio switch-off in energy saving modes, the experiments were designed considering sub-networks of 40 logical nodes each, sampled out of the 139 nodes of the INDRIYA network. Each sub-network consists of randomly selected nodes, with the condition that each node has at least one neighbor. Each sub-network operates for 3 min. After that, a new sub-network is selected. Individual nodes also change their data generation rates (at the application layer). Rate changes occurred with intervals uniformly distributed between 5 and 45 s. Data rates are also uniformly distributed between 5 and 35 Pkts/s. Data packet size is set to 100 bytes, for a total payload of 114 bytes.

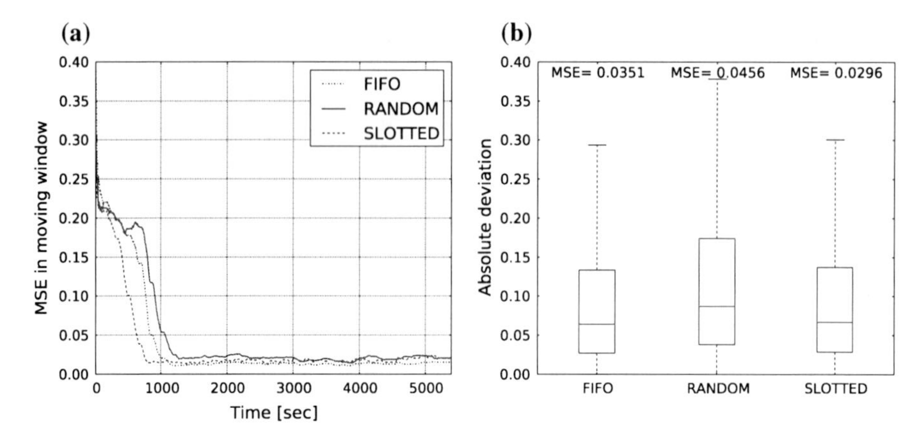

Fig. 1 **a** Moving MSE. **b** Distribution of the errors. Performance of selection strategies for a training set of size 50 samples

Since all data samples are timestamped, we could recreate the batch learning process for each single logical node (represented by different real nodes in each of the sub-networks) and run GPR considering a maximum training set of size 50 and 100 samples comparing the different selection strategies.

As performance metric, we measure *prediction errors*: the difference between measured and predicted values of PRR. These are calculated for each sample immediately after it is processed. We report the mean squared error (MSE) averaged in a moving window. We also report the final MSE value and the boxplot of the distribution of absolute prediction errors. These results are always averaged over all the nodes.

Figures 1 and 2 show the results for training set size of 50 and 100, respectively. From the analysis of the plots we can observe that the SLOTTED and FIFO strategies

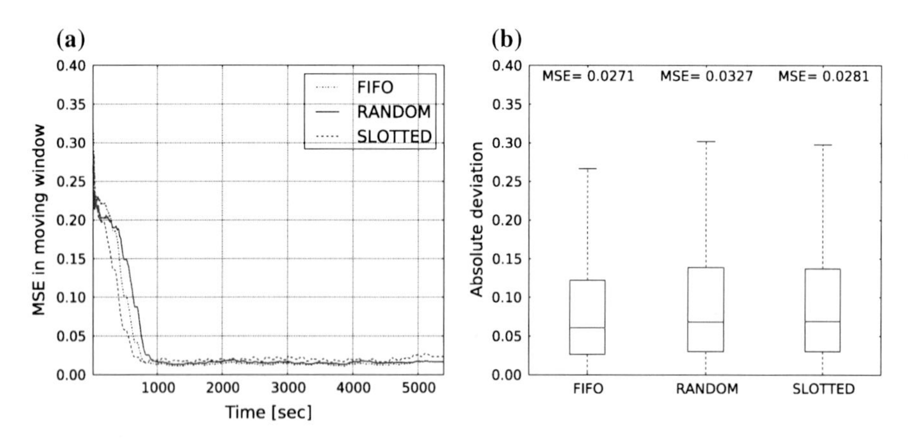

Fig. 2 **a** Moving MSE. **b** Distribution of the errors. Performance of selection strategies for a training set of size 100 samples

provide the best prediction accuracy. Over time, the SLOTTED strategy converges faster (i.e. generates robust models). Although the RANDOM strategy overall does not perform significantly worse, its initial slow convergence suggests that its accuracy could be dramatically reduced in environments that are quite dynamically changing over time.

4 Conclusions and Future Work

We tackled the implementation issues of on-line supervised learning in devices with limited computational resources. We focused on batch learning approaches for processing training data and building a predictor in the context of link quality estimation. Accounting for memory limitations, for buffering training data, we compare three strategies for the on-line selection of the data samples to be kept in memory and used for learning. Results suggest that strategies that keep balanced the set of training samples in terms of ranges of target values provide better accuracy and faster convergence. Future work includes the design of more complex strategies that measure the correlation between the samples composing the training set. The very good prediction accuracy achieved even with small number of samples also motivates the effort to implement the on-line supervised learning algorithms in small, embedded devices such as sensor nodes.

References

1. Vijayakumar, S., D'souza, A., Schaal, S.: Incremental online learning in high dimensions. Neural Comput. **17**, 2602–2634 (2005)
2. Di Caro, G.A., Kudelski, M., Feo, E., Nagi, J., Ahmed, I., Gambardella, L.: On-line supervised learning of link quality estimates in wireless networks. In: Proceedings of the 12th IEEE/IFIP Annual Mediterranean Ad Hoc Networking Workshop, pp. 69–76 (2013)
3. Rasmussen, C.E., Williams, C.K.I.: Gaussian Processes for Machine Learning. MIT Press, Cambridge (2006)
4. Gibbs, M., MacKay, D.J.: Efficient Implementation of Gaussian processes. Technical report, Cavendish Laboratory, Cambridge (1997)
5. Doddavenkatappa, M., Chan, M.C., Ananda, A.: Indriya: A low-cost, 3D wireless sensor network testbed. In: Proceedings of TRIDENTCOM, pp. 302–316 (2011)

Poster Abstract: An Experimental Study of Attacks on the Availability of Glossy

Kasun Hewage and Thiemo Voigt

Abstract Glossy is a reliable and low latency flooding mechanism which makes use of constructive interference. Therefore, it is important to investigate what happens when attacks are mounted on Glossy that try to break constructive interference. In this chapter, we explore the effectiveness of different methods of breaking constructive interference in Glossy. Our results show that Glossy is quite robust to approaches where nodes do not respect the timing constraints necessary to create constructive interference. Changing the packet content, however, has a more tremendous effect on the packet reception rate.

1 Introduction

The Low-Power Wireless Bus (LWB) provides a shared communication bus like infrastructure with a flat network hierarchy [1]. Therefore, no routing protocol is required as nodes can receive packets sent by each other similar to the way of receiving packets from neighborhood nodes. The LWB is built on top of Glossy [2], a reliable and low latency flooding mechanism which has in-built accurate global time synchronization. The network-wide flooding in Glossy and the LWB itself make it possible to support multiple communication patterns such as one-to-many, many-to-one and many-to-many.

The capture effect is a phenomenon that enables the reception of packets despite interference from other wireless signals [3]. However, its efficiency is decreased when concurrent transmissions are increased [4]. In other words, when the node density is

K. Hewage (✉) · T. Voigt
Uppsala University, Uppsala, Sweden
e-mail: kasun.hewage@it.uu.se

T. Voigt
SICS Swedish ICT, Kista, Sweden
e-mail: thiemo@sics.se

K. Langendoen et al. (eds.), *Real-World Wireless Sensor Networks*,
Lecture Notes in Electrical Engineering 281, DOI: 10.1007/978-3-319-03071-5_13,

increased, packet reception is affected negatively since concurrent transmissions may interfere destructively. Glossy uses simultaneous transmissions of the same packet by allowing them to interfere constructively.

As Glossy primarily depends on constructive interference, it is important to investigate what happens when attacks are mounted on Glossy that try to break constructive interference.

In this chapter, we explore the effectiveness of different methods to break constructive interference in Glossy. Our results show that Glossy is quite robust to attacks where the malicious nodes do not respect the timing constraints necessary to create constructive interference. Changing the packet content, however, has a more tremendous effect on the packet reception rate.

2 Background

2.1 Capture Effect and Constructive Interference

The capture effect is the phenomenon associated with packet reception in which the radio is able to receive a packet from one sender despite the simultaneous transmissions from other transmitters. Suppose, two packets from two transmitters in which the strength of first packet is higher than the second packet at the receiver. Assume that the first packet arrives at the receiver earlier than the second packet. If the signal to noise ratio (SNR) of the first packet is higher than a certain threshold, the radio is able to receive the first packet successfully. If the second packet arrives earlier than the packet, the SNR of the second packet falls below the specific threshold since the first packet has a higher signal strength. In such a situation, the second packet is certainly lost. The correct reception of the first packet depends on the radio is detecting the preamble and the Start of Frame Delimiter (SFD) of the packet.

For IEEE 802.15.4 radios, the temporal displacement between two concurrent and identical signals should be less than 0.5 μs in order to correctly receive the packet with a high probability. If a packet is received under such circumstances, this is called constructive interference. The simulations conducted by Ferrari et al. [2] shows that the probability of correct packet detection decreases vastly if the temporal displacement is larger than 0.5 μs. Therefore, generating constructive interference requires tight time synchronization among the radio transmitters.

2.2 Glossy

Glossy uses concurrent transmissions to benefit from constructive interference. The node that starts the transmission is called *initiator* and the other nodes are called *receivers*. Upon reception of a packet from the *initiator*, all *receivers* relay the packet at the same time. The nodes that receive the relayed packet also do the same and the process continues for a number of predefined rounds. In this way, the whole network is flooded within a few milliseconds.

Glossy also provides time synchronization with no additional cost in which the error is below one microsecond. Each packet in Glossy has a 1-byte field called *relay counter*. Initially, this counter is zero and it is incremented at each node before relaying the packet. Upon receiving a packet, a node can determine the number of hops a packet has made. The *slot length* is defined as the duration between the start of two consecutive transmissions of the same packet with the *relay counter*. The *slot length* is a network-wide constant since the packet length is not changed during flooding and locally estimated by the nodes. Therefore, a node can compute the time when the flooding is started, called *reference time*, by the *initiator* based on the *relay counter*. With the knowledge of the *initiator's* clock value and the hop count, other nodes are able to achieve absolute time synchronization.

3 Breaking Constructive Interference

In this section we present the three methods for breaking constructive interference we have used in our experiments.

Delaying the packet relaying (DPR) In order to increase temporal displacement among signals above 0.5 μs, we use *no operation* instructions (NOPs) to delay for triggering the transmission request to the radio. In the MSP430 MCU, NOPs are emulated from other instructions and their execution time can be calculated by using the Digitally Controlled Oscillator frequency of the MCU. Therefore, it is straightforward to compute the number of NOPs required for a delay of above 0.5 μs.

Relaying the packet earlier (RPE) Relaying the packet earlier than the planned time of the transmissions is also a possible way to increase the temporal displacement. In the implementation of Glossy, a certain number of NOPs are used to compensate interrupt serving delay and hardware variations. We adjust these NOPs to relay the packet earlier than the planned time.

Modifying the packet (MP) In addition to increasing temporal displacement, modifying the packet before relaying can also be used to interfere the signals destructively. This is due to the overlapping of different signals. As mentioned in Sect. 2.2, Glossy uses a 1-byte field as the relay counter. Another possible attack is to modify the *relay counter* before relaying.

4 Experimental Results

We conduct our experiments on the FlockLab testbed [5] using Tmote Sky sensor nodes. We select one of the 30 nodes as the *initiator* and one as the malicious node that tries to break constructive interference by using the methods explained in the previous section (DPR, RPE and MP).

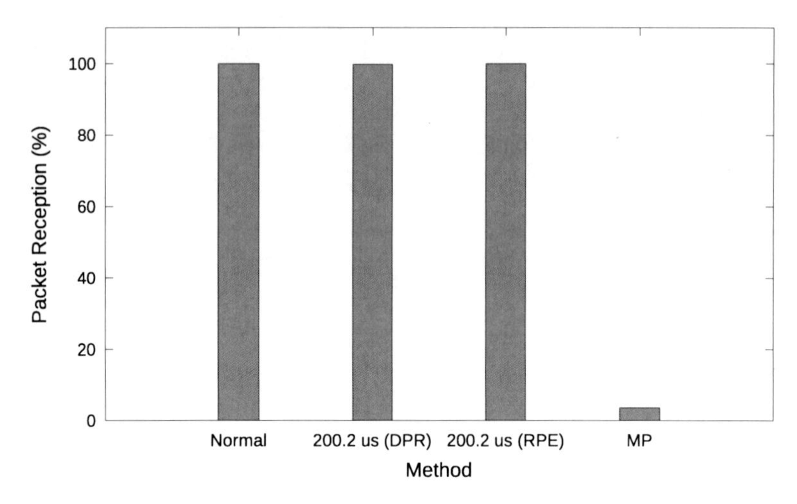

Fig. 1 Packet reception for six selected nodes in the neighborhood of malicious node

The *initiator* periodically starts a Glossy phase flooding the network. The next Glossy phase is scheduled after the end of each Glossy phase. At the *receivers*, the scheduler periodically starts Glossy to receive flooded packets. After the end of each Glossy phase, the skew between node's clock and the *initiator's* clock is estimated by using old and recently estimated *reference times*. The next Glossy phase is scheduled based on the clock skew and the recently estimated *reference time*. In this way, Glossy phases of the *receivers* are always time synchronized with the *initiator* by ensuring the participation of the *receivers* to the flooding. At each node, we log several statistics such as packet reception, CRC failure count, bad-length and bad-header count.

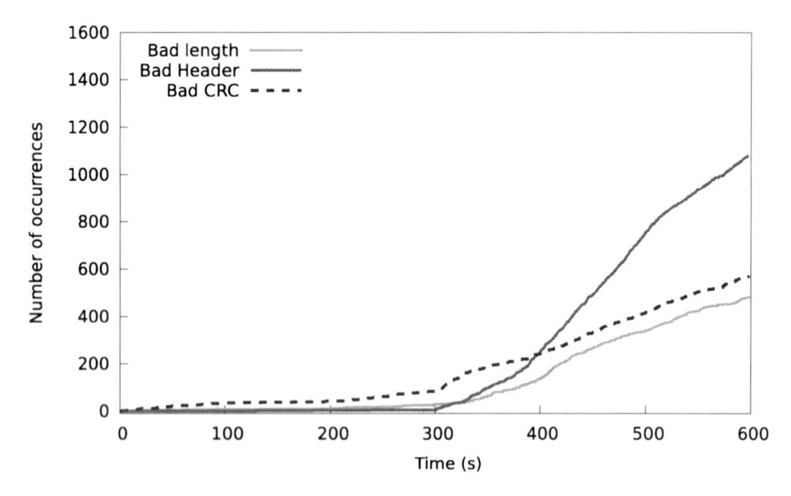

Fig. 2 CRC failures, bad-length and bad-header variation for a selected node (DPR)

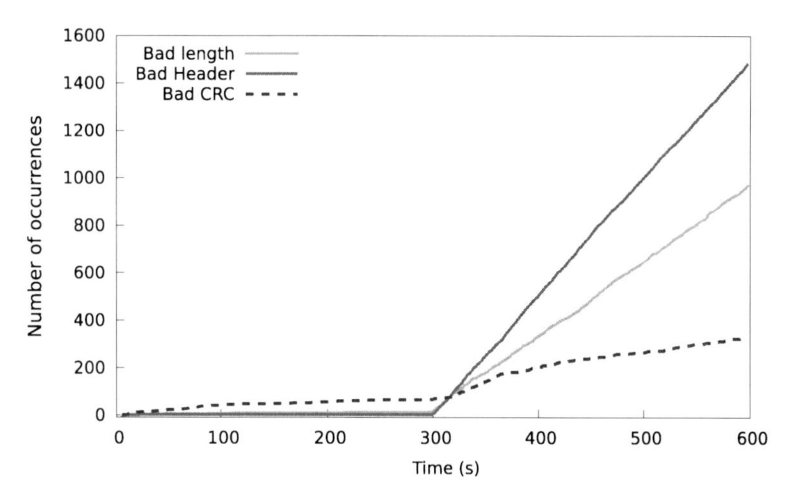

Fig. 3 CRC failures, bad-length and bad-header variation for a selected node (RPE)

Figure 1 shows the packet reception for six selected nodes in the neighborhood of the malicious node. The graph show that Glossy is not affected at all by DPR and RPE. It is important to note that the low packet reception for MP is because Glossy floods become unsynchronized. As a result of modifying the relay counter, nodes incorrectly estimate the *reference time*.

Figures 2 and 3 show the variation of CRC failure count, bad-length and bad-header count over the time for DPR and RPE for a selected node. The graphs show the increase of these errors after the malicious node has been started (after 300 s). Glossy's multiple relays of the same packet make Glossy robust and keep the overall packet loss low despite the attacks.

Acknowledgments We would like to thank Federico Ferrari for clarifications regarding Glossy internals. This work was supported by the WISENET center at Uppsala University.

References

1. Ferrari, F., Zimmerling, M., Mottola, L., Thiele, L.: Low-power wireless bus. In: ACM SenSys '12, New York, NY, USA, ACM 2012
2. Ferrari, F., Zimmerling, M., Thiele, L., Saukh, O.: Efficient network flooding and time synchronization with Glossy. In: ACM/IEEE IPSN, Chicago, IL, USA April 2011
3. Leentvaar, K., Flint, J.: The capture effect in FM receivers. IEEE Trans. Commun. **24**(5), 531–539 (1976)
4. Lu, J., Whitehouse, K.: Flash flooding: exploiting the capture effect for rapid flooding in wireless sensor networks. In: IEEE INFOCOM, Rio de Janeriro, Brazil 2009
5. Lim, R., Ferrari, F., Zimmerling, M., Walser, C., Sommer, P., Beutel, J.: Flocklab: a testbed for distributed, synchronized tracing and profiling of wireless embedded systems. In: ACM/IEEE IPSN, Philadelphia, USA April 2013

Part III
Low-level Components

Node Identification Using Clock Skew

Ibrahim Ethem Bagci and Utz Roedig

Abstract Clocks on wireless sensor nodes experience a natural drift. This clock skew is unique for each node as it depends on the clocks manufacturing characteristics. Clock skew can be used as unique node identifier which is, among other applications, useful for node authentication. We describe how clock skew of a node's clock can be measured directly on a node by utilising the available high precision radio transceiver clock. We detail an implementation of this proposed local clock skew tracking method for the Zolertia Z1 platform. We determine the required sampling effort to accurately determine clock skew. We also discuss how clock skew measurements can be aligned with existing transceiver operations in order to avoid an increase in energy consumption.

1 Introduction

All clocks on wireless sensor network (WSN) platforms experience a natural drift. This drift is unique to a node as it depends on the clock hardware. For example, the drift of a node's real-time clock is defined by unique properties of the used quartz crystal. For most WSN applications clock drift is a nuisance and mechanisms such as time synchronisation protocols are put into place to combat it. However, in this chapter we take advantage of a node's unique clock drift pattern and use it to uniquely identify nodes.

Besides other security requirements it must be possible to authenticate nodes and sensor data provided by them. For example, a sink node must be able to verify that data is provided by genuine nodes and not by an adversary. Classical cryptographic

I. E. Bagci (✉) · U. Roedig
School of Computing and Communications, Lancaster University, Lancaster, UK
e-mail: i.bagci@lancaster.ac.uk

U. Roedig
e-mail: u.roedig@lancaster.ac.uk

K. Langendoen et al. (eds.), *Real-World Wireless Sensor Networks*,
Lecture Notes in Electrical Engineering 281, DOI: 10.1007/978-3-319-03071-5_14,

methods can be used to implement authentication. In this case shared keys are used to identify nodes. As keys may become compromised (i.e. someone obtaines a copy) which would allow an adversary to impersonate a node, methods have been developed which bind authentication to a node's hardware. For example, a crypto chip such as the Atmel ATSHA204 [1] can be included in a node's design which holds cryptographic material for authentication. In this case an attacker must obtain the crypto chip in order to impersonate a node. However, crypto chips are expensive and require an additional component to be integrated in the node design. Thus, we use an already present hardware characteristic to derive a unique node identification; we use a node's unique clock skew characteristic for identification.

In addition to the outlined security application, clock skew based identification is useful for other tasks. For example, when commisioning nodes unique identifiers such as node IDs and communication addresses must be determined. Clock skew can also be used in this broader context to generate unique identifiers.

Clock skew has been previously used as means of node identification. For example, Kohno et al. [2] have shown that clock skew is unique and can be used to identify classical PCs in the Internet. Uddin et al. [3] have shown that this method can be used in principal in the context of wireless sensor networks. Existing work determines clock skew by comparing clocks on separate nodes (or nodes and a sink). For this process it is necessary to distribute time stamps over the underlying communication network and a *constant* network delay is required. In a WSN context this is an impractical requirement as duty cycled communication induces large network delay variances.

In our work we move away from this limitation of existing work and we measure clock skew locally on nodes. We discuss how this can be achieved in general, describe an implementation of this method and provide a detailed analysis describing clock sampling requirements and achievable quality. More specifically, the contributions of this chapter are:

- *Local Skew Determination:* We describe how the clock skew of a node's crystal-based real-time clock can be measured locally using the high precision clock available on modern transceivers to create unique node identifiers. An implementation of this method for the Zolertia Z1 platform using Contiki is described.
- *Analysis of Sampling Requirements:* The achievable quality of the local clock skew determination is compared with the quality of state of the art distributed methods. The dependency of clock sampling effort and skew calculation quality is analysed in detail. In this context it is also shown how clock sampling can be aligned with general transceiver operation in order to avoid a transceiver duty cycle increase.

The remaining chapter is organised as follows. The Sect. 2 describes related work. Section 3 defines the term clock skew formally and methods for clock skew calculation and analysis are discussed. Section 4 discusses remote clock skew determination while Sect. 5 describes local clock skew determination. In Sect. 6 we discuss findings and describe future directions of research work.

2 Related Work

Clock skew is the deviation of a clock from the true time. Fingerprinting devices using clock skew is carried out by comparing frequencies of two clocks, one of them generally assumed to represent the true time.

Kohno et al. [2] has shown that clock skew of devices can be measured to fingerprint devices. It is shown that the clock skew of each device is unique and stays fairly consistent over time.

Zander et al. [4] improved clock skew measurement by applying a technique called synchronized sampling. They demonstrated that synchronization of samples reduces the quantisation error and hence improves skew determination quality.

Jana et al. [5] used clock skew to fingerprint wireless devices. The motivation for their work was the detection of fake wireless access points. Their work demonstrates that emitting timestamped beacons with high frequency allows for precise clock skew calculation. According to their observations 50–100 beacons are sufficient to estimate clock skew accurately enough to identify individual nodes.

Arackaparambil et al. [6] demonstrated a clock skew spoofing attack in 802.11 networks by using virtual interfaces. In their work they propose methods to combat clock skew spoofing and propose standardised interfaces which would allow network providers to publish clock skew information.

Huang et al. [7] demonstrated clock skew-based identification in wireless sensor networks in the context of the Flooding Time Synchronization Protocol (FTSP) [8]. FTSP provides coarse estimation of clock skew based on current offset and previous skew (it uses linear regression on the past eight data points).

Uddin et al. [3] demonstrated that sensor nodes have a unique clock skew and that the clock skew of a node can easily be monitored.

Murdoch et al. [9] split skew into two components, a constant and a variable part. The variable part is affected by temperature changes and this effect was used to reveal node identities in the TOR network by influencing CPU load and hence the temperature of devices leading to measurable clock skew changes.

Our work differs from existing approaches as we measure clock skew locally on nodes. We believe that this is a necessary step towards a practical system as variations in communication delays cannot be avoided in any real-world WSN deployment. Furthermore, we show how clock skew measurements fit with energy efficient operations of sensor nodes and investigate the required sampling effort in detail. Existing work with the exception of Huang et al. [7] calculate clock skew offline after a long sequence of samples are collected using a linear programming approach. Huang et al. [7] calculate clock skew online using a linear regression approach which we adopt in our work.

3 Clock Skew

Clock skew could be determined by analysing the drift of a clock C_m with the help of a stable reference clock C_r. However, in a practical setting a stable reference clock is generally not available and it is only possible to monitor one drifting clock with another drifting clock. Hence, a measured clock skew value for a node reflects drift of the measured clock and the used reference clock. Nevertheless, the determined clock skew value is unique and dependant on the hardware characteristics of the clocks used.

3.1 Definition of Clock Skew

The measured clock C_m runs at frequency f_s and the reference clock C_r runs at frequency f_r. To determine clock skew, timestamps of the measured clock and the reference clock are taken periodically. T_1^m and T_1^r are the first timestamps of both clocks taken at the first sample point, T_i^m and T_i^r are timestamps taken at the ith sample point. The elapsed time of the measured clock C_m at the ith sample point is $t_i^m = (T_i^m - T_1^m)/f_m$; the elapsed time at the reference clock C_r is $t_i^r = (T_i^r - T_1^r)/f_r$. The *offset*—the difference between measured and reference clock—at the ith sample point is $o_i = t_i^m - t_i^r$. If we sample N pairs of (t_i^r, o_i) for $i \in \{1, ..., N\}$ and plot these pairs (the so called offset-set), we obtain an approximately linear graph. It is possible to fit a linear function of the form

$$\delta \cdot t_N^r + \varphi \tag{1}$$

to these obtained measurement points. The slope δ of the fitted linear function is called the *clock skew*.

3.2 Clock Skew Determination

There are a number of methods available to fit a linear function. Depending on the exact method used the definition of *clock skew* is also slightly altered. In the literature mainly two methods are used in the context of clock skew calculation which we detail next and use in the remainder of the chapter.

Linear Programming—LP Moon et al. [10] have shown that clock skew can be accurately estimated from a set of samples using a linear programming method. LP finds a line $\delta \cdot t_i^r + \varphi$ that upper bounds the offset-set. The problem constraint of LP is given as

$$\delta \cdot t_i^r + \varphi \geq o_i \tag{2}$$

and the following function is minimized (object function):

$$\frac{1}{N} \cdot \sum_{i=1}^{N} (\delta \cdot t_i^r + \varphi - o_i) \tag{3}$$

LP delivers accurate results but has drawbacks when considering a WSN context. Samples must be collected and stored before the calculation can begin. Furthermore, a relatively complex solver for the LP must be available. If considering to execute the calculation on a resource constrained node storage and calculation requirements may be too excessive.

Linear Regression—LR Maróti et al. [8] estimate clock skew using a form of linear regression in their Flooding Time Synchronisation Protocol (FTSP). The algorithm uses the average elapsed time at the reference clock $\overline{t^r}$ and the average offset $\overline{o}$ up to the ith sample point. Then the skew is estimated by calculating:

$$\delta = (o_i - \overline{o})/(t_i^r - \overline{t^r}) \tag{4}$$

For the FTSP eight sample points are used to estimate the clock skew. We use this algorithm with a varying number of sample points in order to have control over the achievable quality of the skew estimation. Compared to LP the implementation is much simpler and therefore more suitable for usage on resource constrained nodes.

3.3 Clock Skew Quality

In this work we aim to use clock skew to uniquely identify nodes. Hence it is important that clock skew measurements on two nodes can be attributed to the individual nodes. The collision probability (the likelihood that two nodes are seen as the same even though they are different) should be as small as possible.

Clock skew measurement, for example using the previously described LP or LR method, is subject to variation. Thus, as two nodes may have a skew values close to each other, variance of the measured skew may make it hard to clearly attribute measurements to individual nodes.

We use a t-test to compare the means of the measured clock skews of two nodes. The t-test returns a test decision for the null hypothesis that the nodes are the same. The alternative hypothesis is that the two nodes are not the same. The probability value (p-value) returned by the t-test is the probability of wrongly assuming that the two analysed nodes are the same when they are in fact not. Therefore, a small p-value indicates that the means of clock skew measurements of two investigated nodes are unlikely to be the same.

For the experimental evaluations described in later sections we use p-values to describe quality of determined clock skew measurements.

4 Remote Clock Skew Determination

In existing work (for example, [3] and [7]) measured clock C_m and reference clock C_r are located on different nodes in the network. For example, the reference clock is the real-time clock of the sink node and the measured clock is the real-time clock of the node to be identified. This remote clock skew determination is carried out as a node generally provides only one accurate clock in its default configuration which is suitable for skew calculations.

4.1 The Impact of Network Jitter

As shown in Sect. 3, timestamps T_i^m and T_i^r are the ith sample taken at the same point in time. When using remote clock skew determination T_i^m is taken first and the timestamp is transmitted via the network to the reference node which then takes the corresponding time stamp T_i^r. Thus, T_i^r is taken Δ_i after the sample T_i^s. Δ_i is the network delay associated with transmitting the ith timestamp of the measured clock. The clock skew calculation as presented in Sect. 3 is still valid if the network delay is constant ($\Delta_i = \Delta \forall i$). Variations in Δ_i (network jitter) reduce the quality with which the clock skew can be dertermined. In a practical WSN network jitter is high due to the nature of duty cycled communication. For example, when using a protocol such as ContikiMAC [11] the forwarding delay is dependant on when a sender transmit request occurs relatively to the point in time when a receiver node enters its periodic listen phase. Using standard ContikiMAC configuration settings, forwarding delays vary between 0 and 125 ms (ContikiMAC uses channel check rate of 8 Hz). This is too high to measure the characteristics of clock skew.

Existing work uses remote clock skew determination in specific setups where nodes are only one hop distance away and no duty cycling MAC protocols or significant network traffic is present. In such a specific case network jitter is low and remote skew determination is possible. However, in any practical setting this method has its limitations.

4.2 Experimental Evaluation

We use five Z1 nodes from Zolertia [12] to carry out a baseline experiment using remote clock skew determination. We use the results obtained in this experiment as comparison for the local clock skew determination method we introduce later.

The Z1's external 32,768 Hz watch crystal is used for clock skew profiling. We label the nodes from Node 1 to 5 and select Node 2 as the sink node. Nodes are one hop away from a sink node and its external crystal clock is used as reference clock C_r. The nodes run the Contiki operating system and use a NullMAC (no dutycycling

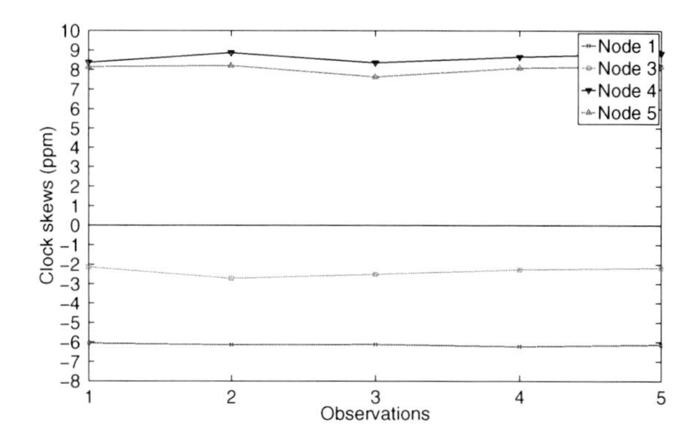

Fig. 1 The measured clock skew using five Zolertia Z1 nodes with remote clock skew determination. Five observations are carried out using 2,500 timestamp samples

MAC is present). No other network traffic than the transmission of timestamps from the four measured nodes is present in the network.

Timestamps T_i^m are transmitted every 4 s from the nodes to the sink. When the sink node receives a timestamp T_i^m it records the corresponding timestamp T_i^r. The total transmitted packet count is 2,500 for every node. We repeat this operation five times (five observations) and estimate the clock skew for each node. The clock skew is estimated using the previous described LP method. This calculation is carried out offline after collection of all timestamp samples. The result is shown in Fig. 1; clock skew is shown in "part per million" (ppm).

Figure 1 shows that nodes can be clearly identified by measuring clock skew. The clock skew is stable enough over several observations. For some nodes it is easier to distinguish them (for example, Node 1 and Node 4 are clearly separate nodes), others have skew values close together (for example, Node 4 and Node 5). However, even though some nodes are close together measured skew values can be clearly attributed to nodes. How clearly nodes can be identified as separate (the quality

Table 1 Obtained p-values describing how clearly nodes can be distinguished from each other node

	Node 1	Node 3	Node 4	Node 5
Node 1	–	$4.25E-06$	$3.19E-08$	$2.36E-08$
Node 3	$4.25E-06$	–	$3.91E-07$	$1.69E-07$
Node 4	$3.19E-08$	$3.91E-07$	–	$3.23E-03$
Node 5	$2.36E-08$	$1.69E-07$	$3.23E-03$	–

The smaller the value the more clearly nodes are distinguishable

Table 2 Obtained p-values when comparing a node with itself

Node 1	Node 3	Node 4	Node 5
$1.14E-01$	$9.43E-01$	$4.49E-01$	$3.54E-01$

Values are larger (2 magnitudes) then the ones shown in Table 1, indicating skews are not distinguisable of the skew values) can be expressed using the previously outlined t-test. Table 1 shows the results of this analysis. To provide some means of judging p-values we provide Table 2. Here, the means of the first two observations of a node are compared with the mean of the third and forth observation of a node; effectively p-values are generated where a node is compared with itself and values are generated that are not distinguishable. As it can be seen, worst case p-values in Table 1 are two orders of magnitude lower then in Table 2. This indicates that the investigated set of nodes in this experiment is clearly uniquely identifiable via measured clock skews.

5 Local Clock Skew Determination

To overcome the previously outlined limitations of remote clock skew determination (i.e. the need of a constant network delay), it would be beneficial to use two local clocks on a node for skew determination.

5.1 Local Clock Sources

Most sensor node platforms provide two clock sources, the crystal-based real-time clock and a processor internal digitally controlled oscillator (DCO). However, these two available local clocks cannot be used for skew determination as the DCO clock has a much lower precision then the real-time clock. Thus, no stable clock skew values can be determined using this setting.

However, most node platforms have a radio transceiver chip which has internally a high precision clock which is necessary for timing of data transmissions. In most cases it is possible to access this resource and use it within the platform for other purposes than transmission and reception of data.

The Zolertia Z1 platform we use for our work provides an 8 MHz clock within the CC2420 radio transceiver. We use the same approach that Pettinato et al. [13] used to make use of the clock source of the radio. The Clear Channel Assessment (CCA) pin of the radio is alternatively configured to output the internal radio clock signal [14]. We connect the radio CCA pin with the external timer sources pin (TBCLK) of the MSP430 processor. This allows us to use the radio clock as alternative clock source. As this clock is of better quality than the crystal based RT clock it is now possible to use these two clocks to determine locally on the node a stable clock skew value.

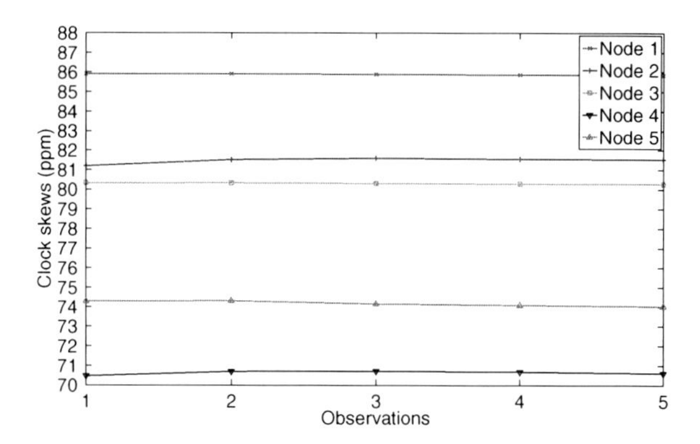

Fig. 2 The measured clock skew using five Zolertia Z1 nodes with local clock skew determination. Five observations are carried out using 600 timestamp samples

5.2 Experimental Evaluation

In this experiment we use the same five Z1 nodes as used for the previously described baseline experiment. The Z1's external 32,768 Hz watch crystal (the measured clock C_m) is used again for clock skew profiling. The CC2420 radio clock source is used as reference clock C_r. Obviously, all nodes have their individual transceiver clock and, thus, for each node skew profile an individual reference clock is used. Timestamps T_i^m and T_i^r are collected with an interval of 1 s. The results determined using the LP method are shown in Fig. 2.

As it can be seen, nodes are clearly identifiable in terms of observed clock skew. The skew values are different to the values obtained remotely as shown in Fig. 1. This has to be expected as different reference clocks are used.

The calculated p-values for the five nodes are shown in Table 3. As can be seen here, p-values are several magnitudes lower compared to the remote skew determination. This means that individual nodes can be distinguished much more clearly. It has to be noted that this significant improvement is achieved even though less sample points (600 compared to 2,500) are taken and a shorter sample durations (1 s compared to 4 s) are used.

We conclude that the local clock skew determination is much better then remote clock skew determination as nodes can be more clearly identified. The collision probability (the likelihood that two nodes are seen as the same even though they are different) is greatly reduced. This is an important factor for the design of security mechanisms.

5.3 Processing Optimisation

So far we have used the LP method to determine clock skew. This method is useful and provides sufficiently accurate results (as previously shown) for local and remote clock skew determination. However, LP is too complex to execute directly on resource constrained nodes. We therefore use the LR method which is computationally more feasible. LR is expected to deliver results of lesser quality compared to the LP method. It is therefore necessary to analyse if quality of the results is still sufficient to distinguish individual node clock skews.

Figure 3 shows a comparison of local skew determination using the LP and LR skew calculation method (The figure also contains lines showing the effect of reducing the sample period from 1 s down to 7.8125 ms; we discuss the effect of reduced sample period in the next section). The corresponding p-values are given in Table 3. Interestingly, the LR method fares sightly better in terms of producing clearly distinguishable clock skew values. This is contrary to what we would have expected. Clearly, LR therefore represents a feasible option for clock skew calculation.

We have implemented the LR method of calculation for the Contiki operating system (LP was executed in Matlab after data had been collected). The run-time complexity of the LR algorithm is $O(n)$. For the calculation of skew using 200 sample points an execution time of 2.01 ms is required which corresponds to an energy consumption of 0.027 mJ.

Table 3 p-values using LP and LR with 10 observations, 600 samples, 1 s sample period

	Method	Node 1	Node 2	Node 3	Node 4	Node 5
Node 1	LP	–	$1.30E-15$	$1.6E-24$	$8.13E-22$	$6.58E-19$
	LR	–	$1.92E-16$	$6.12E-25$	$1.74E-22$	$3.12E-19$
Node 2	LP	$1.30E-15$	–	$1.63E-10$	$8.32E-21$	$1.21E-15$
	LR	$1.92E-16$	–	$1.59E-11$	$8.79E-21$	$1.46E-16$
Node 3	LP	$1.36E-24$	$1.63E-10$	–	$4.13E-20$	$8.29E-17$
	LR	$6.12E-25$	$1.59E-11$	–	$7.01E-21$	$3.98E-17$
Node 4	LP	$8.13E-22$	$8.32E-21$	$4.13E-20$	–	$2.03E-13$
	LR	$1.74E-22$	$8.79E-21$	$7.01E-21$	–	$4.36E-14$
Node 5	LP	$6.58E-19$	$1.21E-15$	$8.29E-17$	$2.03E-13$	–
	LR	$3.12E-19$	$1.46E-16$	$3.98E-17$	$4.36E-14$	–

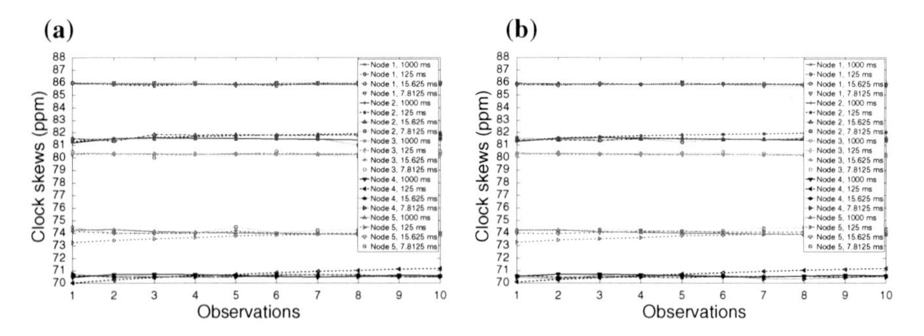

Fig. 3 **a** Clock skews using LP. **b** Clock skews using LR. The measured clock skew of five Zolertia Z1 nodes with remote local clock skew determination. Ten independent observations with 600 timestamp samples are used

5.4 Sampling Optimisation

So far we have demonstrated the feasibility of local clock skew determination using relatively long sampling periods of 1 s and and a relatively large sample size of 600 samples. It is beneficial to reduce as much as possible sample period and sample size. Reducing the sample period and size allows for more energy efficient operations and enables us to obtain skew values faster. When reducing the sample period less time to observe skew is available and resolution of the used clocks can be a limiting factor. When reducing the sample size less points are available to reduce variance of the obtained result.

For local clock skew determination as presented in this chapter the transceiver clock is required. This clock is only available if the transceiver chip is active. A duty cycled MAC protocol will aim to put the transceiver into an energy efficient sleep state for as long as possible and only wake the transceiver for short durations for transmissions and receptions. If short sample periods are feasible it is possible to align communication and clock sampling and no additional transceiver chip wake times must be introduced. For example, many slotted MAC protocols have transceivers periodically on for durations of 10 ms to facilitate reception or transmission.

The required sample size should be as small as possible as this would allow us to obtain a skew measurement fast. If we assume that sample periods are aligned with natural transceiver activity we still have to wait until the transceiver was used sufficiently often enough before a skew measurement can be obtained.

We record 600 samples with four different sampling intervals: 1 s, 125 ms, 15.625 ms and 7.8125 ms. These results are shown for 10 observations in Fig. 3, using LP and LR for analysis. Figure 4a, b shows changes in skew variance of 10 observations when modifying sample size and sample period.

It is clearly visible that skew variations experienced from one measurement to the next are reduced when increasing sample period and/or sample size. The question is what a feasible combination of sample size and sample period is. The answer will depend on the application situation. However, generally it can be assumed that the shortest feasible sampling period should be used as this helps in aligning sampling

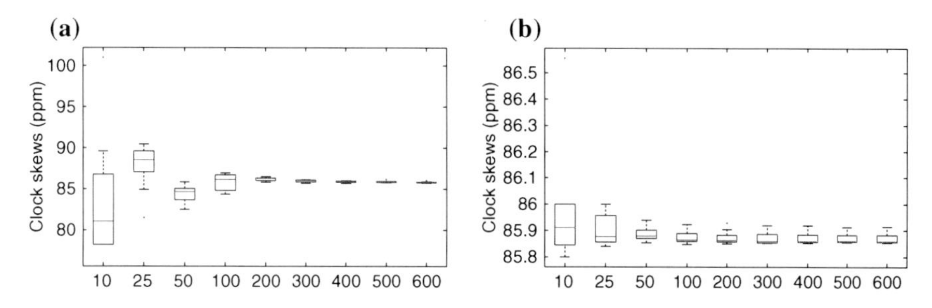

Fig. 4 **a** Node 1 using a 7.8125 ms sample period. **b** Node 1 using a 1 s sample period. Node 1 skew for different sample sizes and sample period of 7.8125 ms and 1 s

Table 4 *p*-values using LR with 10 observations, 200 samples, 7.8125 ms sample period

	Node 1	Node 2	Node 3	Node 4	Node 5
Node 1	–	$6.66E-10$	$6.14E-12$	$3.14E-16$	$3.37E-15$
Node 2	$6.66E-10$	–	$1.21E-04$	$3.25E-12$	$2.08E-12$
Node 3	$6.14E-12$	$1.21E-04$	–	$3.79E-12$	$1.92E-11$
Node 4	$3.14E-16$	$3.25E-12$	$3.79E-12$	–	$1.69E-09$
Node 5	$3.37E-15$	$2.08E-12$	$1.92E-11$	$1.69E-09$	–

with general transceiver activity. Then, the number of samples should be reduced up to a point a sufficient quality (expressed as *p*-values) of skew calculation is ensured. For example, if we assume that a sample period of 7.8125 ms is chosen and a sample size of 200 is chosen we obtain *p*-values as shown in Table 4. These *p*-values are better than the *p*-values obtained via remote skew determination. Thus, with these settings local skew detection can replace remote skew detection without a loss in skew quality.

6 Conclusion

A node's unique clock skew can be used for node identification purposes. It is useful to use this approach because node identification is bound to the hardware and no additional components have to be incorporated in the node design. We have demonstrated that clock skew can be determined reliably locally on nodes. Existing methods rely on jitter free network communication which is unachievable in most practical WSN deployments. Thus, the presented work takes an important step towards practical clock skew identification.

We have shown that clock skew of a node's RT clock can be determined locally using the transceiver clock present in most WSN systems. We have shown that locally determined skew values for nodes can be as unique and distinguishable as skew values determined in a distributed fashion. A sample period of 7.8125 ms and a sample size of 200 is sufficient to determine clock skew locally with the same quality as remotely with a sample period of 4 s and a sample size of 2,500. Also, the possible short sample period of 7.8125 ms allows us to take clock skew measurements during times the transciever is active for communication. Additional transceiver active periods do not have to be scheduled to achieve local clock skew determination.

Initial experiments have shown that the measured clock skew depends on temperature. A node would need to be profiled in terms of skew over the expected temperature range. Skew values would have to be transmitted together with a temperature reading in order to allow identification in deployments with varying temperature.

References

1. Atmel: Atmel ATSHA204 datasheet. http://www.atmel.com/Images/Atmel-8740-CryptoAuth-ATSHA204-Datasheet.pdf (2012). Accessed Mar 2012
2. Kohno, T., Broido, A., Claffy, K.: Remote physical device fingerprinting. IEEE Trans. Dependable Secure Comput. **2**(2), 93–108 (2005)
3. Uddin, M., Castelluccia, C.: Toward clock skew based wireless sensor node services. In: Wireless Internet Conference (WICON), 2010 The 5th Annual ICST, pp. 1–9 (2010)
4. Zander, S., Murdoch, S.J.: An improved clock-skew measurement technique for revealing hidden services. In: Proceedings of the 17th Conference on Security Symposium. SS'08, pp. 211–225. USENIX Association, Berkeley (2008)
5. Jana, S., Kasera, S.K.: On fast and accurate detection of unauthorized wireless access points using clock skews. In: Proceedings of the 14th ACM International Conference on Mobile Computing and Networking. MobiCom '08, pp. 104–115. ACM, New York (2008)
6. Arackaparambil, C., Bratus, S., Shubina, A., Kotz, D.: On the reliability of wireless fingerprinting using clock skews. In: Proceedings of the Third ACM Conference on Wireless Network Security. WiSec '10, pp. 169–174. ACM, New York (2010)
7. Huang, D.J., Teng, W.C., Wang, C.Y., Huang, H.Y., Hellerstein, J.: Clock skew based node identification in wireless sensor networks. In: Global Telecommunications Conference, 2008. IEEE GLOBECOM 2008, pp. 1–5. IEEE (2008)
8. Maróti, M., Kusy, B., Simon, G., Lédeczi, A.: The flooding time synchronization protocol. In: Proceedings of the 2nd International Conference on Embedded Networked Sensor Systems. SenSys '04, pp. 39–49. ACM, New York (2004)
9. Murdoch, S.J.: Hot or not: revealing hidden services by their clock skew. In: Proceedings of the 13th ACM Conference on Computer and Communications Security. CCS '06, pp. 27–36. ACM, New York (2006)
10. Moon, S., Skelly, P., Towsley, D.: Estimation and removal of clock skew from network delay measurements. In: INFOCOM '99. Eighteenth Annual Joint Conference of the IEEE Computer and Communications Societies. Proceedings, vol. 1, pp. 227–234. IEEE (1999)
11. Dunkels, A.: The ContikiMAC Radio Duty Cycling Protocol. Technical Report T2011:13, Swedish Institute of Computer Science. Dec 2011
12. Zolertia: Zolertia Z1 datasheet. http://zolertia.com/sites/default/files/Zolertia-Z1-Datasheet.pdf (2010). Accessed Mar 2010
13. Pettinato, P., Wirstrm, N., Eriksson, J., Voigt, T.: Multi-channel two-way time of flight sensor network ranging. In: Picco, G., Heinzelman, W. (eds.) Wireless Sensor Networks. Lecture Notes in Computer Science, vol. 7158, pp. 163–178. Springer, Berlin (2012)
14. Texas Instruments: 2.4 GHz IEEE 802.15.4 / ZigBee-Ready RF Transceiver (Rev. C). http://www.ti.com/lit/ds/symlink/cc2420.pdf (2013). Accessed Mar 2013

MagoNode: Advantages of RF Front-ends in Wireless Sensor Networks

Mario Paoli, Antonio Lo Russo, Ugo Maria Colesanti and Andrea Vitaletti

Abstract This chapter introduces the MagoNode: a new low-power wireless device for Wireless Sensor Networks operating in the ISM 2.4 Ghz band. This platform is based on a highly efficient RF front-end that greatly improves RF performance, in terms of radio range and sensibility, still limiting energy consumption. Indeed, the device outperforms other existing amplified platforms available on the market and is comparable to most commonly known unamplified motes. Moreover, the MagoNode is tailored to operate in various countries with different wireless regulations. Our platform is also TinyOS-compatible and supports the Contiki operating system.

1 Introduction

Most of the modern hardware platforms dedicated to Wireless Sensor Networks (WSN) use 802.15.4 compliant transceivers operating in the ISM 2.4 GHz band [1, 2]. Advantages are the world-wide availability of the 2.4 Ghz band, higher data rates with respect to other ISM bands (433 Mhz, 868/915 Mhz), almost no regulatory restrictions related to the duty cycle [3, 4] and compliance to ZigBee, that is a widespread standard solution for remote monitoring and control applications [5]. Unfortunately,

M. Paoli (✉)
Dipartimento di Informatica, Sapienza Università di Roma, Rome, Italy
e-mail: paoli@dis.uniroma1.it; mariopaoli85@gmail.com

A. Lo Russo · U. M. Colesanti · A. Vitaletti
Dipartimento di Ingegneria Informatica, Automatica e Gestionale Antonio Ruberti, Sapienza Università di Roma, Rome, Italy
e-mail: lorusso@dis.uniroma1.it

U. M. Colesanti
e-mail: colesanti@dis.uniroma1.it

A. Vitaletti
e-mail: vitaletti@dis.uniroma1.it

K. Langendoen et al. (eds.), *Real-World Wireless Sensor Networks*,
Lecture Notes in Electrical Engineering 281, DOI: 10.1007/978-3-319-03071-5_15,

those advantages come at the expense of a lower transceiver sensibility and higher propagation losses due to the higher transmission frequency which translates in a shorter radio range, especially for indoor applications.

In several practical scenarios, the limited radio range of 802.15.4 transceivers requires the presence of many additional relay nodes to guarantee network connectivity. Common assumptions in theoretical works on WSNs are that the density of the network can be arbitrarily high to guarantee the robustness of the network and to improve the quality of collected data. However, in many practical cases those assumptions are questionable. In our experience in monitoring critical infrastructures, a limited number of very accurate and costly transducers need to be deployed in given positions potentially far from each other. This implies several relaying nodes to be placed in order to guarantee a reliable and robust network. However, a high number of relay nodes implies higher hardware and deployment costs as well as an increase in energy costs due to more complex coordination and management activities. As an example, in our WSN test-bed of the Roman B1 line underground construction site, a 700 m tunnel has been monitored by our network for months (from November 2011 until June 2012) [6]. To guarantee wireless connectivity and an acceptable link quality, nodes were about 40–50 m apart. However, experts required the monitoring of only four sections of the tunnel spaced from 100 to 400 m one from the other. Hence, at the end of the deployment, the tunnel was monitored by 32 nodes: one acting as sink, 23 as relay nodes and only eight as nodes interfaced with transducers.

To overcome this issue, some recent hardware solutions for WSNs have started to propose built-in RF front-ends [7, 8]. An RF front-end is a combination of a Power Amplifier (PA) and a Low Noise Amplifier (LNA) that enhances the transmission and the reception of the transceiver, at the expense of an energy cost overhead. However, the always increasing efficiency of transceivers combined with the constantly lowering power consumption of RF front-ends make this kind of solution more and more convenient.

Contribution of the chapter In this chapter we present the MagoNode platform that has been developed at Dipartimento di Ingegneria Informatica, Automatica e Gestionale "Antonio Ruberti" [9] of Sapienza University of Rome in cooperation with the spin-off WSense s.r.l. [10]. This platform takes advantage of the hardware improvements discussed earlier, and provides outstanding radio performance, still keeping the advantages of standard 2.4 Ghz transceivers. In the remainder of this chapter we briefly describe commercial RF front-ends features and the most recent nodes equipped with such devices. We present our platform with its extension boards and compare it with other amplified motes. We show how our platform is more flexible and has better performance in terms of power consumption, than motes equipped with a RF front-end and, at the same time, it is still competitive with respect to traditional unamplified nodes. Finally, in the last section we compare the MagoNode platform and the IRIS while running an indoor test-bed, to evaluate the advantages of an RF front-end approach.

2 RF Front-ends and Platforms

2.1 Front-ends Requirements

As mentioned in the previous section, an RF front-end embeds a low noise amplifier (LNA) to improve the receiver sensitivity and a power amplifier (PA) to increase the output power. In addition to several research chapters focused on low-power 2.4 Ghz RF front-ends for WSNs [11, 12], a wide range of devices are now available at most common RF hardware manufacturers. We took care of RF front-ends suitable for WSN applications, relying on the following parameters:

- **Supply voltage** most wireless nodes operates in the 1.8–3.6 V range in order to be powered by 2×AA 1.5 V batteries, hence, we expect a front-end to be as close as possible to this range.
- **RX Gain** and **RX Current** keeping in mind that idle listening is the most common source of energy waste in a WSN [13], the RX current overhead of the front-end must be as low as possible, still guaranteeing a good RX gain.
- **TX Gain** and **TX Current** the TX power output is regulated by local organizations such as ETSI in Europe and FCC in North America. In particular, ETSI limits the maximum output power to +20 dBm [3, par. 4.3.2.1] but the actual output power is limited by the power spectral density specification in [3, par. 4.3.2.2] which reduces the limit to approximately +10 dBm [14, 15]. Similarly, the FCC has a maximum output power limit of +30 dBm [4], however, the stricter radiation emission limit, difficult to meet without expensive filtering, reduce the actual power to +18–20 dBm [14, 15]. It is important to note that the TX current overhead is typically much higher than the RX one but, at the same time, in WSNs the time spent in TX and RX is much lower than the time spent in idle listening [13].
- **Noise figure** each RF front-end amplifies the RF signal together with RF noise. The lowest is the noise figure, the better the transceiver will behave.

In Tables 1 and 2 we have compared several RF front-ends products available on the market based on the aforementioned parameters. The values indicated in those tables have been taken or derived from the available products data-sheets. As it can be depicted, there is no evident winning solution: the best RX performance is given by CC2590 and T7024, but the former has a high noise figure and is limited to a maximum power output (P_{out}) of +14 dBm in TX while the latter, despite the good current/gain ratio, has a high RX current consumption. The best TX performance is given by RF6555 which is weak in RX due to a high current drain. Skyworks SE2431L seems to represent a good balance between TX and RX modes and natively supports antenna diversity option.

However, using an RF front-end tailored for 20 dBm power output becomes counterproductive when the P_{out} is limited to 10 dBm. In fact, using a front-end with lower maximum P_{out} like the CC2590 or SE2438T allows to reach current consumption @10 dBm that is less than half with respect to the other solutions. Based on the

Table 1 Front-ends characteristics (part 1)

Brand	Model	Current RX (mA)	Gain (dB)	Current TX (mA) @10dBm	Current TX (mA) @20dBm	Noise Fig. (dB)
Skyworks	SE2431L	5	12.5	50	115	2
Skyworks	SE2438T	5.5	10.5	20	N/A	3.5
Skyworks	SKY65344	7	10	N/A	105	2.2
Skyworks	SKY65352	7	10	72	110	2
RFMD	RF6555	8	11	60	90	3
RFMD	RF6575	8	13	90	170	2.5
Atmel	T7024	8	16	50	110	2.1
TI	CC2590	3.4	11.4	17	N/A	4.6
TI	CC2591	3.4	11	50	105	4.6
RFAxis	RFX2401C	10	12	N/A	100	2.4
RFAxis	RFX2411	8	12	N/A	110	2.5

Table 2 Front-ends characteristics (part 2)

Brand	Model	Max P_{out} (dBm)	Voltage (V)	Comments
Skyworks	SE2431L	23	2–3.6	Antenna div.
Skyworks	SE2438T	16	2–3.6	Antenna div.
Skyworks	SKY65344	20	2.7–3.6	
Skyworks	SKY65352	20	2.7–3.6	
RFMD	RF6555	20	2–3.6	Antenna div.
RFMD	RF6575	22	3–3.6	Antenna div.
Atmel	T7024	23	2.7–4.6	
TI	CC2590	14	2–3.6	Footprint CC2591
TI	CC2591	22	2–3.6	Footprint CC2590
RFAxis	RFX2401C	22	2–3.6	
RFAxis	RFX2411	21	2–3.6	

previous observation, it is possible to design efficient solutions fitted for different market areas and regulations. For the same reason, Texas Instruments developed CC2590 and CC2591 as interchangeable devices that share the same footprint and components.

2.2 Competitors: Sensor Nodes with an RF Front-end

In this section we briefly introduce motes for Wireless Sensor Networks featuring RF front-ends.

The AdvanticSYS CM3300 node is equipped with a separated PA and LNA solution, the SiGe PA2423L and Sirenza SGL-0622Z respectively [16]. The CM3300 features 120 mA current consumption for transmission @20 dBm and 30 mA current consumption in RX giving 23 dB gain and is TelosB compatible, hence, it supports TinyOS [17] and Contiki [18] operating systems. However, as opposed to other existing solutions, the CM3300 is designed to run from mains since the 5 mA sleep current does not allow it to be battery-powered for long lasting deployments.

The Atmel ZigBit Amp ATZB-A24-UFL (ZigBit) [7] is an OEM module handling Atmel's 1281v MCU, AT86RF230 transceiver and a Skyworks (formerly SiGe) SE2431L RF front-end. The ZigBit accepts 3–3.6 V input voltage, it features +20 dBm output power and 12.5 dB of RX Gain. A TinyOS porting for the ZigBit mote has been provided by [19].

The Dresden Elektronic deRFmega128-22M12 (deRFmega) is the most recent wireless OEM module available on the market [8]. It is based on Atmel's Atmega128 RFA1 MCU/Transceiver System-on-chip bundle and the Skyworks SE2431L front-end. The device accepts from 2 to 3.6 V in input and has the following characteristics (derived from the user manual [20]): 18.5 mA RX power consumption with 12.5 dB gain, 143 mA for TX@20 dBm and 59 mA for TX@10 dBm. The deRFMega also supports Contiki and TinyOS operating systems.

3 MagoNode Platform

While the integration with an RF front-end has obviously represented the predominant part of the platform development, several other aspects, closer to commonly known wireless sensor nodes requirements, have been taken into account during the MagoNode design.

3.1 Platform Design

Controller selection We looked toward a cheap, low-power microcontroller and transceiver bundle, possibly with software support for existing WSN operating systems (TinyOS, Contiki) and with all commonly used digital and analog interfaces. Atmel's Atmega128RFA1 (RFA1) met all our requirements being a 8-bit, 16 Mhz System-On-Chip (SoC) 802.15.4 compliant with 128 KB of ROM, 16 KB of RAM and with outstanding transceiver performance: 103.5 dB link budget, a current consumption of 12.5 mA in RX and 14.5 mA in TX @+3.5 dBm. The SoC characteristic of the RFA1 chip reduces both size and bill of materials and keeps the overall platform costs very low.

The RFA1 supports Atmel's BitCloud ZigBee stack and is natively supported by TinyOS since release 2.1.2 thanks to the UCMote mini platform [2]. Also Contiki OS supports the RFA1.

RF front-end We based our RF front-end selection on two fundamentals observations: first of all, we expect idle listening to represent the predominant energy cost on a wireless node, hence, it becomes important to pay as little RX overhead as

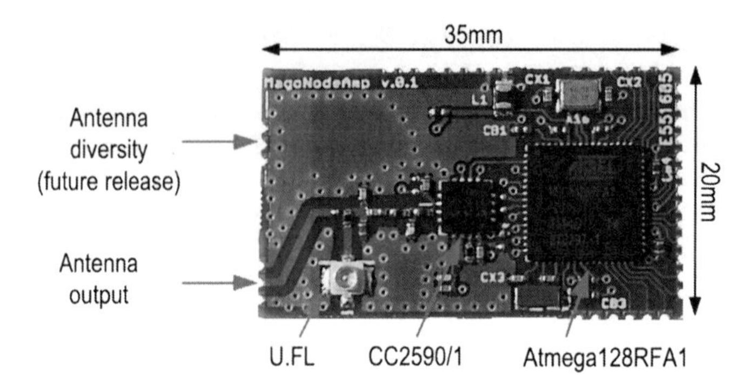

Fig. 1 The MagoNode OEM platform

possible when adding an RF front-end. Secondly, as already mentioned in Sect. 2.1, local regulations considerably limit the maximum output power. Thus, it makes nonsense to use a +20 dBm designed front-end in Europe while it would be limiting to use a +10 dBm designed front-end in North America. At the same time, it is very unlikely that the same hardware used in one geographical area will move in a completely different region. Based on these observations and having a look back to Table 1, we observe how CC2590 and CC2591 represent an optimal solution for our requirements. In particular, both front-ends have the lowest RX current consumption still guaranteeing good RX gains and, most importantly, they both share the same footprint, making them interchangeable over the same PCB design.

PCB design The MagoNode platform is designed as a 4-layer OEM board, i.e., a castellated PCB solderable on bigger application-specific boards. The reason of such choice with respect to more academic-like solutions such as TelosB or MicaZ, resides in a higher flexibility in board design for end-users applications. The MagoNode OEM board is 35 mm long and 20 mm large (Fig. 1). It already holds the RF filtering section designed following [21] and a U.FL connector that allows the connection of an external antenna plug. The board has 39 castellated pins featuring most of RFA1 capabilities, including 2 UART interfaces, 1 SPI, 1 I^2C, more than 20 GPIOs and 4 10-bit ADC channels. One pin can be enabled as antenna output by moving a 0 Ohm resistor. Finally, the board footprint has been designed to enable antenna-diversity in future releases.

3.2 Boards

As as a first step, we integrated the MagoNode in two different boards design (Fig. 2). The first one is a common academic-like board, namely MNA-Board, featuring a 2×AA battery holder, a power switch, three debug leds, an RP-SMA connector, a 51 pin Hirose expansion connector and, optionally, a 2 MB flash chip. The board is 32 mm large and 55 mm long and, as almost all commonly known wireless sensor

Fig. 2 The current-loop interface (*left*) and the MNA-board (*right*)

nodes, allows quick prototyping and debugging as much as easy deployment. The second board is an application-specific sensor-board which acts as an interface for 4–20 mA Current-Loop sensors. The 55 × 55 mm board accepts 5–30 V input voltage, provides a digitally controlled power switch for the current loop sensor and has three 24-bit ADC channels with precision resistors for current measurements. The board is housed in an IP56 box and uses an external antenna plugged to the U.FL connector of the MagoNode.

3.3 Radio Current Consumptions: a Performance Comparison

To evaluate the performance of the MagoNode platform, in terms of current consumption, we have run two sets of tests. On the one side we compared the European (CC2590) and North American (CC2591) versions of the MagoNode platform with respect to similar state-of-art amplified wireless nodes mentioned in Sect. 2.2, i.e., ZigBit and deRFMega. On the other side, we compared the CC2590 version of the MagoNode with respect to common unamplified motes like IRIS, TelosB and the Atmel Atmega128RFA1-EK1 which shares the same MCU/Transceiver bundle with the MagoNode but has not an RF front-end. Actually, we want to demonstrate not only the competitiveness of our solution compared to other amplified devices, but also that the energy overhead introduced by the MagoNode is low enough to get compared with unamplified motes.

We programmed each device with the RadioCountsToLeds application available in the official TinyOS 2.1.2 distribution. With a Rigol DM3068 precision multimeter we sampled the current consumption with a period of 100 μs. Finally, we calculated the average current consumption of each platform in reception (RX), transmission (TX) and idle listening (IDLE). Note that for the deRFMega platform, we derived those values from [20, Fig. 9 and 10] since the device was not available in our lab.

Figure 3 shows the current consumption of amplified nodes with the radio in TX@20dBm, in TX@10dBm, in RX and in IDLE. At highest output power, performance of the MagoNode CC2591 overwhelms those of deRFmega and Zigbit

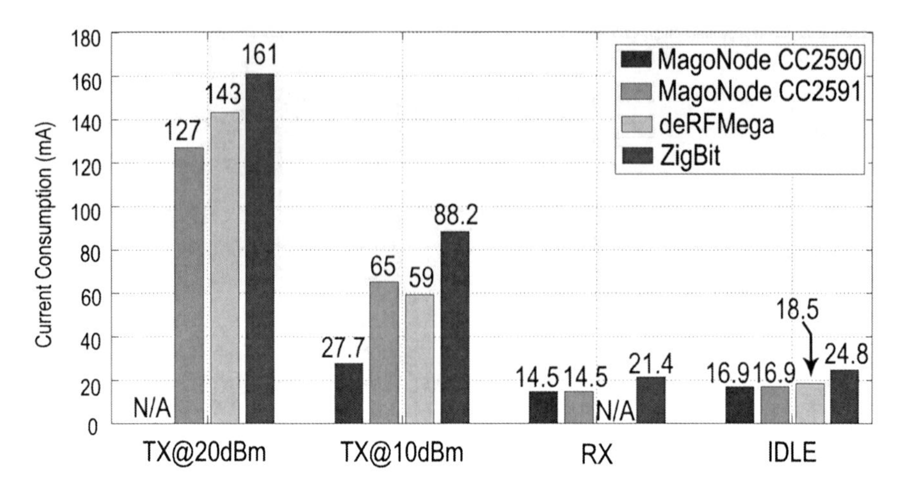

Fig. 3 Current consumption comparison between the MagoNode and other amplified motes. The CC2590 has a maximum $P_{out} = 14\,\text{dBm}$ and deRFMega datasheet does not provide the RX current consumption

by 12 and 21 % respectively. At 10 dBm the MagoNode CC2590, which is optimized for this power setting, has a current consumption than is less than half with respect to others. This validates the choice we made to equip the MagoNode with two interchangeable front-ends that provide the best performance at 10 dBm and 20 dBm respectively. In addition, the IDLE performance of CC2590 and that one of CC2591 beat that of deRFMega by 9 % and that of ZigBit by 32 %. Similar results arise in RX between MagoNode and ZigBit platforms, while it was not possible to compare the deRFMega since the actual RX consumption was not reported in [20].

Figure 4 shows the current consumption of unamplified nodes and that of the MagoNode CC2590 with the radio in TX, in RX and in IDLE. We observe how the MagoNode CC2590 has a sensibly higher current consumption in TX than those of IRIS@3 dBm (29 %), TelosB@0 dBm (45 %) and EK1@3,5 dBm (90 %). However, both RX and IDLE consumption of the CC2590 are lower than IRIS and TelosB by 19 and 35 % respectively. Of course, being the EK1 similar to a non-amplified version of the MagoNode, even in these two radio states it drains less current by 28 % in IDLE and 12 % in RX. Anyway, using the MagoNode in a real multi-hop scenario, thanks to the extended radio range and enhanced link quality guaranteed by the RF front-end, results in a reduction of the network traffic due to a lower number of re-transmissions and forwards. This implies that the measured higher current consumption of the MagoNode is mitigated by an overall lower number of expected transmissions, as we will clarify in the next section.

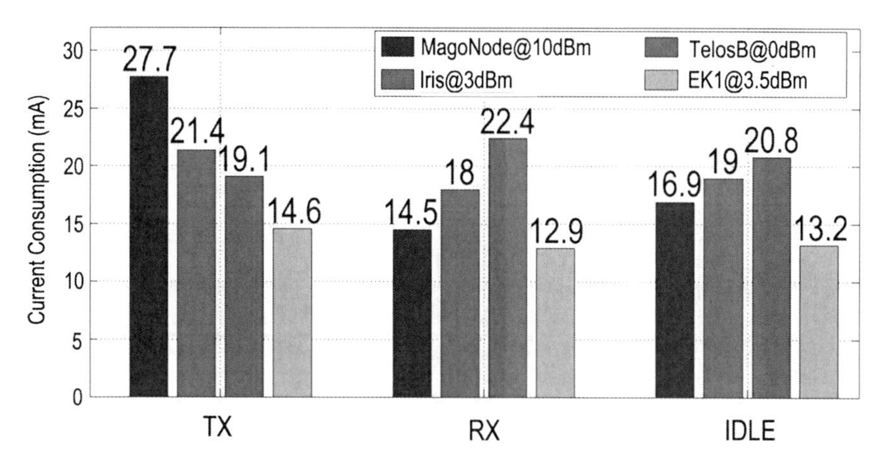

Fig. 4 Current consumption comparison between the MagoNode and unamplified motes. EK1 is comparable to a MagoNode without the RF front-end

4 Testbed

In the previous section we observed how the MagoNode platform outperforms the other amplified solutions. At the same time we also showed how the performance of the CC2590 version of the MagoNode are close to the other unamplified nodes. These results lead us to think that, in a real multi-hop context, where the actual energy consumption is driven by re-transmissions and hop count, the benefits introduced by the front-end we chose for the MagoNode platform (CC2590) can further reduce the energy overhead with regard to unamplified nodes.

4.1 Setup

We deployed the two networks in the basement of our department: the first one made of MagoNode CC2950 and the second one composed by IRIS motes (Fig. 5). The test-beds ran in two different days and for the same amount of time. Like many indoor tests the location is characterized by obstacles that shrink the radio range of the nodes. In the specific, the basement of our building features 70 cm walls thickness and metal doors.

Each test-bed is made of 20 nodes powered by 2 × 1.5 V AA alkaline batteries. We used the Collection Tree Protocol (CTP) [22, 23] as routing layer and Low Power Listening (LPL) [24] as MAC, both are implemented in the TinyOS 2.1.2 distribution. In particular, we exploited the RFXlink library which enforces a common radio stack for both platforms. We tuned CTP and LPL following parameters listed in Table 3,

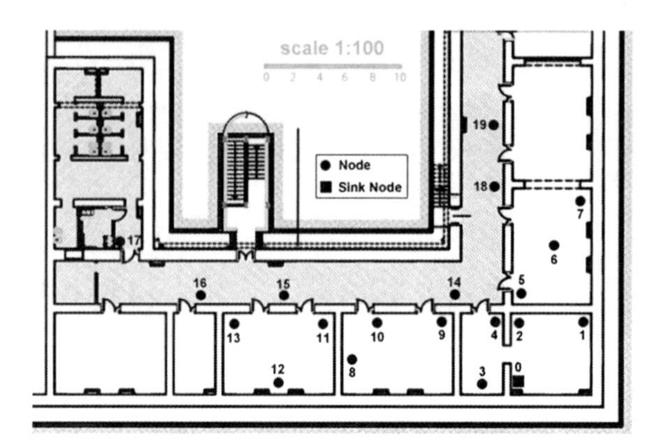

Fig. 5 Test-bed topology

Table 3 Test-bed parameters

Parameters	Value	Parameters	Value
Inter-packet interval	300 s	CTP SENDDONE OK OFFSET	500 ms
Battery	2 × 1.5 V	CTP SENDDONE OK WINDOW	250 ms
Packet size	103 bytes	CTP LOOPY OFFSET	4 s
Run time	24 h	CTP LOOPY WINDOW	4 s
Nodes number	20	CTP NO ROUTE RETRY	20 s
LPL wakeup period	500 ms	CTP MIN INTERVAL	1 s
LPL listen	10 ms	CTP MAX INTERVAL	500 s
LPL delay after receive	20 ms	RF Channel	11

please refer to [23] and [24] for parameters details. Each test-bed ran for 24 h, each node transmitted a data packet toward the sink (Node 0) with an Inter-Packet Interval (IPI) of 5 min, using channel 11 of 802.15.4. We activated the TrafficMonitorLayer of the RFXlink library which enables detailed network statistics related to the radio activity such as the overall radio active time, the time spent in transmission and reception, the overall number of bytes transmitted and received. All these values are carried by each transmitted data packet which has a length of 103 bytes.

4.2 Metrics

We compared both test-beds in terms of network reliability, generated traffic and the subsequent power consumption. We evaluated the network reliability statistics through the following metrics:

- **Data Delivery Ratio (DDR)** this is the ratio between data packets transmitted by each nodes and actually received by the sink.

- **Total Parent Chgs** CTP automatically selects the parent of a node based on the link reliability. This metric represents the overall number of parent changes of the whole network.

On the other side, the generated traffic statistics were calculated exploiting the following metrics:

- **Total ReTx** CTP features automatic re-transmission of unacknowledged data packets. The overall count is aggregated in this metric.
- **Average Time Has Lived (THL)** each data packet carries the hop count represented by the THL metric provided by CTP. The average THL is then an average routing-level network diameter.
- **Total Fwd** this metric represents the overall number of data packets that have been forwarded over the whole multi-hop network.

Finally, we computed the power consumption statistics, derived from TrafficMonitorLayer, exploiting the following metrics:

- **Total CC [TX, RX, IDLE]** these metrics point out the overall current consumption of the network in mJ when nodes are respectively in TX, RX and IDLE state.
- **Total TS [TX, RX, IDLE]** these metrics point out the overall time spent by nodes in TX, RX and IDLE state.

4.3 Results

In Table 4, we can see how the reliability is guaranteed by CTP since DDR is equal to 100 % for both networks. However, to do this, the MagoNode network requires a much lower quantity of generated traffic in terms of data re-transmission, forwarded packets and average THL. This is due to the higher transmit power provided by the RF front-end, that extends the radio range of the nodes and strengthens the reliability of the channels. This entails a minor use of the radio that leads to a lower overall power consumption. Indeed, as we can observe from Table 4, the sum of TX, RX and IDLE time of the MagoNode network is sensibly lower with regard to IRIS network, as is the current consumption. In particular, while consumption of the MagoNode network in both TX and RX are close to the IRIS ones, in IDLE, the MagoNode network performance is lower by about 20 %. Note that the higher TX time of the IRIS platform also affects the IDLE time. This is caused by the LPL transmission mechanism which consists in a burst of packets interleaved by backoff and ack-wait periods. During those periods the transceiver is in idle listening state. We also notice how the IDLE power consumption of both networks is more than 90 % of the total one, hence, it represents by far the predominant consumption when comparing networks performance. This observation validates the design choice made in Sect. 3.1 that keeps the RX/IDLE overhead as low as possible.

Table 4 Test-bed results

	MagoNode	IRIS	$\Delta(\%)$
Avg DDR (%)	100	100	0
Tot Parent Chgs	38	114	−66
Tot ReTx	319	1393	−77
Avg THL	1.55	2.14	−28
Tot Fwd	3395	6804	−50
Tot CC TX (mJ)	130379	132655	−1.7
Tot CC RX (mJ)	37973	39391	−3.5
Tot CC IDLE (mJ)	1713799	2139719	−20
Tot TS TX(s)	1569	2066	−24
Tot TS RX (s)	873	729	+2
Tot TS IDLE (s)	33803	37538	−10

Δs are calculated with respect to IRIS values

5 Conclusions

The MagoNode platform outperforms other RF front-end platforms in terms of energy consumption. In particular, we explained how a more flexible platform, that could be equipped with the proper front-end depending on the specific geographical context, helps to keep the current consumption low in many application scenarios. We also evaluated the MagoNode and other unamplified platforms showing that current consumption in TX, RX and IDLE are comparable.

The obvious advantages introduced by an RF front-end such as extended radio range and more reliable links, allows the MagoNode to better fit to many real scenarios, like structural health monitoring, where traditional unamplified motes requires several additional relay nodes to guarantee the appropriate network connectivity. In addition, we compared the MagoNode with the IRIS mote in a real multi-hop indoor scenario, keeping the same topology for both networks. Interestingly, even in this case, the results show that the advantages introduced by the RF front-end of the MagoNode help to contain the overall energy consumption by lowering the overall network traffic.

References

1. Iris, TelosB, Micaz motes. http://www.memsic.com/wireless-sensor-networks
2. UCMote mini. http://www.ucmote.com
3. Etsi, EN 300 328 V1.8.1 final draft (2012–04). http://www.etsi.org
4. Fcc, CFR 47. https://www.fcc.gov/
5. ZigBee Alliance. http://www.zigbee.org
6. GENESI Project Website. http://genesi.di.uniroma1.it
7. ZigBit ATZB-A24-UFL/U0 datasheet. http://www.atmel.com
8. Dresden Elektronik OEM modules deRFmega. http://www.dresden-elektronik.de

9. Dipartimento di Ingegneria Informatica, Automatica e Gestionale "Antonio Ruberti". http://www.dis.uniroma1.it
10. WSense s.r.l. http://www.wsense.it
11. Hao, Z., Zhiqun, L., Meng, Z., Gang, C.: A 2.4 GHz low-IF RF frontend for wireless sensor networks. In: ICMMT 2010, China on May 8–11, 2010
12. Didioui, A., Bernier, C., Morche, D., Sentieys, O.: Impact of RF front-end nonlinearity on WSN communications. In: ISWCS, 2012
13. Demirkol, I., Ersoy, C., Alagöz, F.: MAC protocols for wireless sensor networks: a survey. In: IEEE Communications Magazine, April (2006)
14. Texas Instruments Application Note AN086. http://www.ti.com
15. Jennic support website. http://www.jennic.com/support/solutions/00002
16. AdvanticSYS CM3300 datasheet. http://www.advanticsys.com
17. TinyOS. http://www.tinyos.net
18. Contiki. http://www.contiki-os.org
19. Huber, R., Fahrni, T.: Porting the ZigBit 900 Platform to TinyOS. Feb 2009
20. User Manual Radio Modules V1.1, http://www.dresden-elektronik.de
21. Texas Instruments Application Note AN098. http://www.ti.com
22. Gnawali, O., Fonseca, R., Jamieson, K., Moss, D., Levis, P.: Collection tree protocol. In: SenSys'09, Nov 4–6, Berkeley, CA, USA (2009)
23. Colesanti, U., Santini, S.: The collection tree protocol for the castalia wireless sensor networks simulator. Technical Report Nr. 729, ETH Zurich, June, 2011
24. Moss, D., Hui, J., Klues, K.: Low power listening - TEP105

MIMOSA, a Highly Sensitive and Accurate Power Measurement Technique for Low-Power Systems

Markus Buschhoff, Christian Günter and Olaf Spinczyk

Abstract In the area of wireless sensor networks, mobile computing systems and other battery-driven computing platforms it is an important task to reduce the system's power consumption, as the usability of such a system is tightly coupled to the duration of a battery cycle. A basic task for reducing energy consumption is the creation of hardware and software energy models to find and understand existing power saving potentials. This often requires a most accurate measurement of both the energy and the timing behavior of such a system. In this chapter we will show that, especially for sub-milliampere applications such as sensor network nodes, commonly used power measurement techniques do not perform well. We introduce MIMOSA, a low-cost measurement device that uses a fast voltage regulator and analog integration circuits to overcome most of the issues existing in other approaches.

1 Introduction

To derive energy models from existing hardware and software solutions, extensive measurement efforts must be made, since manufacturers of hardware components often do not declare the energy-related behavior of their products in detail. In many data sheets, the typical values given for digital circuits are the peak power, the general

"**M**essgerät zur **i**ntegrativen **M**essung **o**hne **S**pannungs**a**bfall", German for "Measurement device for integrative measurements without voltage drop".

M. Buschhoff (✉) · C. Günter · O. Spinczyk
Technische Universität Dortmund, 44227 Dortmund, Germany
e-mail: markus.buschhoff@tu-dortmund.de

C. Günter
e-mail: christan.guenter@tu-dortmund.de

O. Spinczyk
e-mail: olaf.spinczyk@tu-dortmund.de

K. Langendoen et al. (eds.), *Real-World Wireless Sensor Networks*,
Lecture Notes in Electrical Engineering 281, DOI: 10.1007/978-3-319-03071-5_16,

power dissipation and the idle power. Often, these values are just peak values for a certain worst-case setup or for a whole product family. Usually, reliable power values for all states of a component are not given. Thus, deriving a model for a complete system of components, which includes a number of CPU cores, memory, I/O hardware, radio- and wired communication devices and a software architecture with multiple implementation layers, requires a sophisticated measurement setup and benchmark utilization.

Central part of such a measurement setup is a digital power-meter, which allows for controlling and sensing different states of a benchmark, while recording the power consumption of a device. An example application may be the evaluation of a peripheral device's energy consumption, including the CPU utilization by the device driver, as shown in Sect. 6. As I/O often is done during a short-time *interrupt service routine* (ISR) within the driver, the measurement equipment needs a sufficient time resolution and electrical current sensitivity. Both can be achieved with standard lab devices, like the Powerscale STD probe system, in situations where the electrical current running through the *device under test* (DUT) is well above several milliamperes.

Nowadays, low-power systems like Texas Intruments' MSP-430 MCU family perform in the sub-milliampere scale. This demands for new approaches in measurement, because small currents in this scale are difficult to sense, especially when there are sporadic peaks and mixed signals from other consumers in the periphery of the MCU. Several manufacturers of lab equipment offer very exact *direct current* (DC) measurement devices, but working solutions for non-DC situations are rare and expensive.

In the following we introduce MIMOSA, a measurement technique which was developed for exploring the power consumption of low-power embedded systems. The MIMOSA device consists of a prototype hardware implementation that allows for measuring the energy consumption of low-current devices with a good time-precision and a guarantee for the acquisition of short energy peaks that is far beyond the abilities of common current-sampling circuits.

This chapter is structured as follows: In the next section, we explain some necessary basics on power measurement technology and discuss the basic ideas behind MIMOSA. Afterwards, we compare existing approaches in Sect. 3. More technical details of the MIMOSA prototype are presented in Sect. 4. In the evaluation section we show how MIMOSA performs in terms of linearity, signal-to-noise distance and in direct comparison to Hitex Powerscale. Finally, we show a practical example where MIMOSA was used to derive an energy model in Sect. 6.

2 Basics

A basic explanation of some terms and concepts of electronic engineering shall help to understand the challenges and approaches described later in this chapter.

At first, the term *power consumption* (measured in Watts, [W]) here describes the current (I) running through a system while a given *constant* supply voltage (V_{sup}) is applied.

$$P(t) = V_{sup} \cdot I(t) \tag{1}$$

Energy consumption, sometimes also referred to as *work*, given in *Joule* [J] or *Watt-Seconds* [Ws], describes the power consumed over a given time.

$$E = \int P(t)\,\mathrm{d}t = V_{sup} \int I(t)\,\mathrm{d}t \tag{2}$$

As power and energy hardly can be measured directly, instead voltage, current and time are to be measured, so that the power- and energy consumption can be calculated.

But, electrical current can only be measured indirectly, as physics allows only for sensing the effects of electrical current, which are electric- and magnetic fields, and a voltage drop over a conductor. The voltage drop is described by the famous Ohm's law (3), which states that the voltage drop V over a conductor with an electrical resistance R is proportional to the electrical current I flowing through the conductor.

$$V = R \cdot I \tag{3}$$

Measuring the voltage drop is the most relevant technique as it is easy to achieve. The current to be measured is led through a high-precision ohmic resistor (R), often referred to as *shunt*. The voltage drop at this resistor is measured and the current and energy can be calculated. This approach has some drawbacks, because the resistor must be dimensioned in regard to two contrary goals: The electrical resistance must be as small as possible to reduce the influence on the measured system, since a big shunt would cause a big voltage drop, leading to a big energy dissipation within the resistor. This energy loss could even lead to malfunctions in the DUT. On the other hand, the shunt also needs to be chosen as big as possible to achieve a considerable voltage drop that can be measured with good precision. The common approach to solve this problem is to choose a very small high-precision shunt, as well as a sensitive (and complex) measurement technology which utilizes signal amplifiers for sensing the resulting small voltage drop. Nonetheless, the small voltage drop still has an unfavorable *signal-to-noise ratio* (SNR) leading to inaccurate measurements.

Additionally, the shunt must be chosen in regard to the measurement scale. As an example, we take a 3 V driven system that should be measured by a maximum voltage drop of 1 %. With a maximum current drain of 50 mA this would require a shunt of 0.6 Ω, while the same voltage drop at 50 μA would require a 600 Ω shunt. This shows that the shunt-selection depends on the energy consumption of DUT. If it needs more energy than expected by the measurement device, the voltage drop can become too high, leading to improper operation of the DUT.

Manufacturers of more sophisticated devices try to circumvent these issues by using an automated shunt selection circuit. As shunt-switching takes a certain amount of time, those devices typically have problems with rapidly changing loads as they might appear in computer systems. Frequently changing between idle-states and active, high-power peripheral I/O, like radio transmissions, might cause the measurement device to continuously switch between shunts, resulting in unreliable output.

Another issue that arises with common measurement devices is the sampling problem: The signal representation of a sampled curve is only valid for frequencies below half of the sample rate (*Nyquist-Shannon sampling theorem*). This means, that for a device with a sampling frequency of 100 kHz, all signals that are close to or shorter than 20 μs might be misrepresented. Figure 1a illustrates how short activity peaks, as they occur in computing devices, might cause a big error in the results.

This led to the following two key ideas for the construction of the MIMOSA device:

Voltage Regulation: The MIMOSA prototype uses a different technique for stabilizing the voltage drop. It utilizes a fast voltage-regulating circuit to compensate it, allowing a high shunt resistance, high voltage drops, and thus removing the need for highly-sensitive measurement technologies. This implies that MIMOSA is not only a measuring device, but also the supply voltage source for the DUT.

Integration: MIMOSA uses a set of analog integration circuits to deal with short peaks. In fact, MIMOSA "calculates" the integral of the shunt's voltage drop on the *analog* side of the circuit by using a well-known setup of operational amplifiers and capacitors. The integral calculation is done for each sampling period.

That means, in contrast to common digitizing circuits, which sample a "snapshot" of the shunt's voltage-drop, our approach is to sample the integrated energy value that was calculated between the samples. As integral calculation is implemented as an analog circuit, we can show that information loss is much lower than by the usual snapshot-sampling. Thus, the output of MIMOSA is not the electrical current measured via small voltage drops, but the energy consumption over the whole last sampling period measured by voltage drops having a high signal-to-noise distance.

Equation (4) shows the mathematical basement of a typical sample-and-connect measurement. For each snapshot sample m at a time t_m over an interval $t_s = t_m - t_{m-1}$ (which is constant), with a sample value of V_m, a linear connection between the measurements is assumed. This is depicted as a gray, triangular function in Fig. 1a. The equation calculates the area under this function for each sample interval, which is $\frac{1}{2}(V_m + V_{m-1}) * t_s$, and sums it up for all samples. It is obvious that the resulting

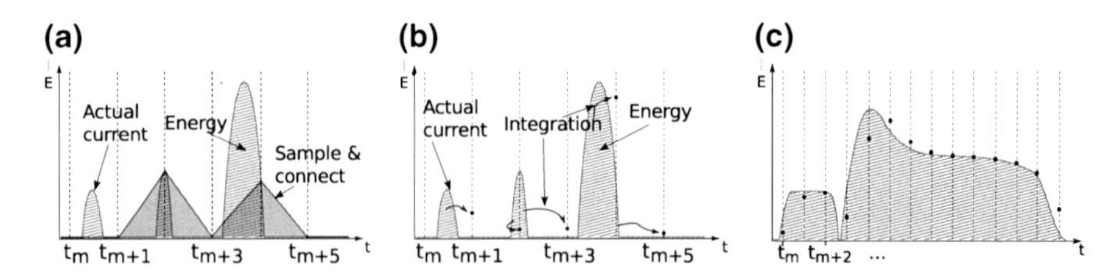

Fig. 1 Signal capturing in typical sampling approaches and MIMOSA. **a** Sample and connect for short peaks showing under and over-represented signals. **b** MIMOSA integrates all signals in an analog circuit. No peaks are lost. **c** Signal shape representation for wide signals in MIMOSA

energy value can only be correct, if the actual signal resembles to the assumed linear connection of the snapshots. This can only be the case for signals that do not have high frequency portions.

$$
\begin{aligned}
E &= V_{sup} \sum_{m=1}^{n} t_s \frac{V_m + V_{m-1}}{2} \cdot \frac{1}{R} \\
&= \frac{t_s \cdot V_{sup}}{2R} \sum_{m=1}^{n} (V_m + V_{m-1})
\end{aligned} \tag{4}
$$

As the MIMOSA approach calculates the integral over each sample period on the analog side of the circuit, no assumptions on the signal shape between two measurements needs to be made. The result of MIMOSA measurements can be calculated as:

$$
\begin{aligned}
E &= V_{sup} \sum_{m=1}^{n} \int_{t_{m-1}}^{t_m} \frac{V(t)}{R} \, \mathrm{d}t \\
&= V_{sup} \int_{t_0}^{t_n} \frac{V(t)}{R} \, \mathrm{d}t \\
&= V_{sup} \int_{t_0}^{t_n} I(t) \, \mathrm{d}t
\end{aligned} \tag{5}
$$

As (5) is equivalent to (2), MIMOSAs accuracy only depends on the quality of the analog integration circuits and the signal-to-noise ratio. Theoretically, it has an infinite vertical resolution. Figure 1b illustrates the MIMOSA results as black dots. The value of each sample is the integral of the last sample-interval (the hatched area under the actual signal curve).

Another important aspect is the shape of a signal. Unlike other sampling circuits, MIMOSA acquires correct energy values for peaks that occur between the samples, but it is not able to show the shape of that peaks. But, for signals significantly longer than a sampling period, MIMOSA still can represent the signal shape quite well, as illustrated in Fig. 1c.

3 Related Work

3.1 Measurement Approaches

An overview of energy measurement methods in the field of embedded systems and network technology is given by Nakutis [1] and Hergenröder et al. [2].

In the field of wireless sensor networks, Jiang et al. proposed SPOT [3]. SPOT uses a differential amplifier for sensing the voltage drop at a low-resistance shunt and a *voltage-to-frequency conversion* (VFC) to overcome limits in the measurement range. The drawback of this method is the noise at a low ohmic shunt and that a counter is necessary to determine the peaks of the VFC's output signal. This means, the resolution is restricted by the frequency used for reading the counter. If it is read with a high frequency, the system has a low resolution. Vice versa, if a high resolution is necessary, only rare measurements are possible.

Another approach is shown by Konstantakos et al. [4]. Here, a *current mirror* instead of a shunt is used to duplicate the input current. Afterwards, a capacitor is charged to integrate over that current. Konstantakos et al. use a method which is closely related to *coulomb counting* and SPOT for determining the input current: Whenever the capacitor reaches a certain charge, it triggers a counter and gets discharged. Thus, it suffers from the same drawback as the SPOT method: The resolution scales with the frequency of the counter readings.

Other, commercially available approaches range from a diversity of power measurement and battery-management chips to high-end lab measurement devices. Typical integrated circuits are the ADE7753 [5], DS2438 [6] (sampling and digital integration) and MCP3906 (VFC) [7]. These solutions lack the necessary dynamic range and resolution, since they are usually just advanced ADCs.

In the middle field of the device range are full-fledged measurement solutions like the HiTex Powerscale system. The Powerscale probe technology automatically switches between shunts to adapt the measurement range. Powerscale was designed and advertised for enabling the user to visualize the power consumption of software routines in embedded systems [8]. Powerscale was awarded with the Embedded World Award 2010 (category "Tools"), and was elected as one of the products of the year 2011 (category "Embedded Design") by the readers of the German "Elektronik" magazine. This is the reason why we compare MIMOSA to the most sensitive and sophisticated of the Powerscale probes: The so called "ACM probe", with a measurement range from 200 nA to 500 mA.

The far end of the device-range is covered by professional lab equipment, like the Agilent P-Series power meters [9], which use high sampling rates and very precise, highly sensitive and well shielded circuits, but are also difficult to handle and cost intensive.

So, to the best of our knowledge, only the high end of the measurement systems might be able to meet the special requirements needed to create accurate energy models of wireless sensor networks or other low-power embedded systems.

4 MIMOSA Technology

4.1 Overview

Figure 2 shows a block diagram of the MIMOSA technology. On the left, the current is converted to a voltage drop over a high ohmic shunt. The resulting voltage drop is compensated by a voltage regulator, which in itself acts as the voltage supply for the DUT.

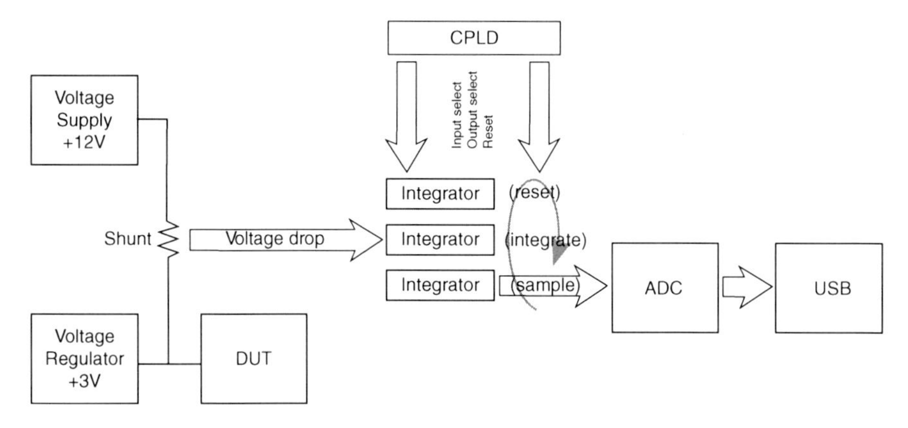

Fig. 2 Overview of the MIMOSA prototype

The voltage drop is accumulated in one of three analog integrators. To avoid gaps during the measurement, three integrators work in a pipeline setup: While one integrator is working for the active sample interval, the output of the last sample interval is send to a an *analog to digital converter* (ADC) by another one. At the same time, the third integrator, which held the results of the second-last interval, is reset to zero. After each sampling interval, the integrators switch their function. The whole process is controlled by a *Complex Programmable Logic Device* (CPLD).

The last part of the signal chain is a data interface, currently implemented as a USB connection. The data interface could be exchanged to allow a broad range of applications, e.g., using an I^2C interface for *in-situ* measurements, as proposed by Jiang et al. [3].

4.2 Voltage Drop Compensation

As MIMOSA can use a high-ohmic shunt to create a considerable voltage drop, a compensational circuit is necessary to allow for fault-free operation of the DUT. For this purpose, Fig. 3a shows how an *operational amplifier* (OpAmp) was used to copy a reference voltage to its output connections. The reference voltage is generated by an integrated high-precision reference-voltage circuit. It must be equal to the required supply voltage of the DUT. As the OpAmp works in a negative feedback loop, it will automatically compensate for any voltage drop at the output node. A downstream power-transistor, which is part of the feedback-loop, allows for higher electrical currents without overloading the OpAmp.

The shown circuit has some important properties for this application:

- The output voltage is almost constant. Its uniformity depends on the *slew rate* of the OpAmp and the downstream transistor.

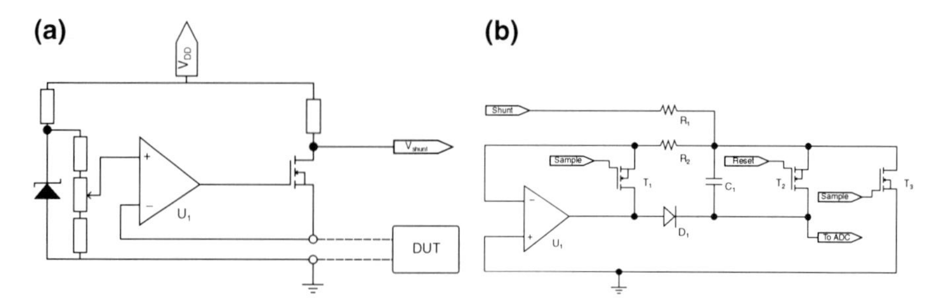

Fig. 3 Simplified schematics of the MIMOSA circuits. **a** Voltage drop compensation. **b** Analog integrator

- The upper voltage of the shunt and its electrical resistance can be freely chosen within the abilities of the active components. So the voltage drop is highly scalable over a broad voltage range.
- The current for the DUT is completely delivered through the shunt branch.
- A drawback of this approach is that the capacitive load must be kept low. However, to achieve good current measurements, a low capacitance is always necessary.

4.3 Analog Integration

Analog integral calculation is a typical application for OpAmps. The simplified schematic of one of the three MIMOSA integration circuits is shown in Fig. 3b. The principle is simple: A capacitor (C_1) is charged with a current I_C proportional to the shunt's voltage drop V_{shunt} (6). The capacitor's charge Q_C is the integral of the incoming current over time (7).

To allow for an electrical current independent of the capacitor's charge, the OpAmp's negative input acts as *virtual ground*. To do so, the OpAmp will continuously compensate the capacitor's charge by lowering its (negative) output voltage, which by that must be proportional to the capacitor's charge. As shown in Equation (8), the output of the integrator thus is proportional to the integral of the shunt's voltage drop.

$$I_C = \frac{V_{shunt}}{R_1} \tag{6}$$

$$Q_C = \int_t I_C(t)\,\mathrm{d}t \tag{7}$$

$$\begin{aligned} V_{out} &= -\frac{Q_C}{C_1} = -\frac{1}{C_1}\int_t I_C(t)\,\mathrm{d}t \\ &= -\frac{1}{R_1 \cdot C_1}\int_t V_{shunt}(t)\,\mathrm{d}t \end{aligned} \tag{8}$$

For MIMOSA, three transistor switches have been inserted into the integrator circuit, which are controlled by the CPLD. When all transistors are off, the circuit is integrating. By activating the *Sample* signal (Transistors T_1 and T_3), the integration is stopped and the output voltage is stable. In this mode, the output voltage can be sampled safely by the downstream ADC.

After sampling, the integrator needs to be reset. This is done using the *Reset* signal on transistor T_2, which short-circuits capacitor C_1 until it is completely discharged.

As stated before, MIMOSA uses three integrators to allow sampling the output of one integrator, while the second is working and the third is being reset.

5 Evaluation

To evaluate the performance of MIMOSA, we have set up a series of experiments to yield reproducible conclusions on the accuracy of the system.

In the first experiment, ten precision resistors were used sequentially to generate a DC load. For each resistor, 300,000 values were taken at a sample rate of 100 kHz, so that each integrator was used 100,000 times. Figure 4a shows the results of this experiment. Each value is denoted by a cross. Since the noise is very low, the crosses appear as a vertical bar. Figure 4b shows that the noise is within a range of ± 3 nJ. The distance of the median from the regression straight is never above 0.3 nJ.

For showing the accuracy of the integrators, we generated peaks smaller than the sample time of 10 μs using a signal generator.[1] The signal generator was set to a rectangular signal with a frequency of 300 kHz, so that 3.33 peaks occurred between two samples. By changing the setting for the duty-cycle, which is the ratio of the high and the low level time of the signal, the pulse duration and thus the output energy of the signal was altered. Figure 5a shows that pulses even smaller than 200 ns were

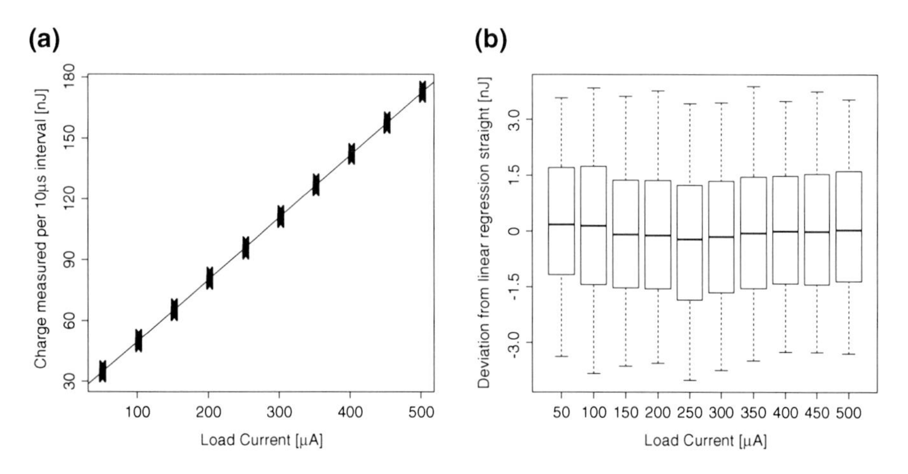

Fig. 4 DC measurement with 10 precision resistors. **a** DC linearity. **b** Noise

[1] PeakTech 4030

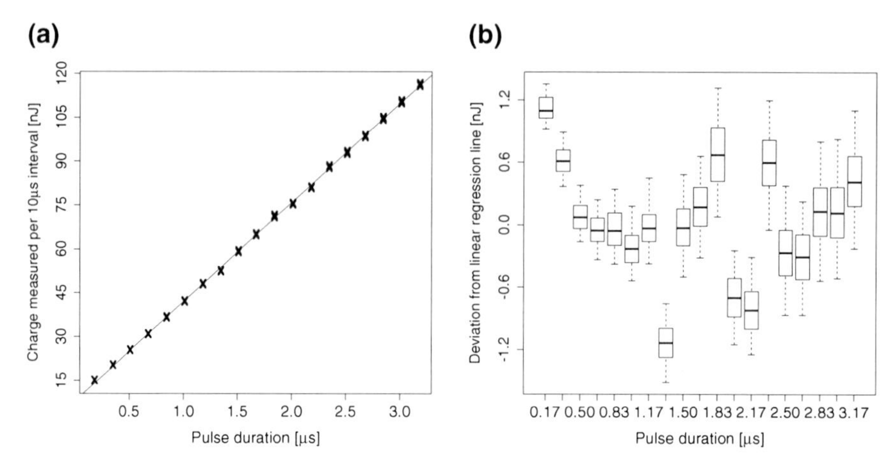

Fig. 5 Integration of peaks smaller than the sample period (10 μs). **a** Linearity. **b** Noise

sensed by the analog integrators and the system behaved linear and showed nearly no noise throughout the whole measurement. In this measurement we show how MIMOSA behaves in the mA range, so that the noise becomes very low (below 1.5 nJ, Fig. 5b).

To evaluate the signal shape in comparison to other devices, we measured a 20 kHz rectangular signal on both MIMOSA and Powerscale with the ACM probe. Both devices sample with a frequency of 100 kHz, so both devices had 5 samples per period, which is close to the limit where sampling can represent a good signal form. Figure 6a shows the resulting curve for both measurements. While MIMOSA is still

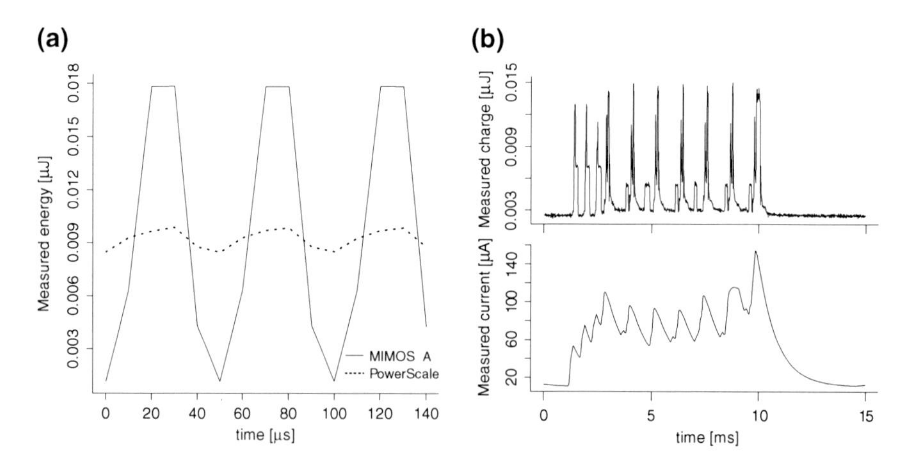

Fig. 6 MIMOSA versus Powerscale. **a** MIMOSA versus Powerscale: Signal shape comparison of a 20 kHz rectangular signal. **b** *Top* MIMOSA; *Bottom* Powerscale

able to show a slightly distorted rectangular signal due to the mathematically correct integration at the flanks, the signal cannot be recognized as rectangular any more using Powerscale.

6 Practical Application

As MIMOSA's main purpose is the mapping of energy-related behavior of a compound system created of hardware and software, we will show an example on how this can be accomplished. The DUT in this example will be a TI eZ430-Chronos watch. This device is an MSP-430 based embedded system comprised of a plenty of sensors and a wireless communication interface.

In our example, we want to measure the power consumption of the CMA-3000 acceleration sensor within the watch, including the software stack and the inter-chip communication that is necessary to retrieve acceleration measurements.

In fact, the motivation for a more sophisticated measurement method arose when we started the experiments for a modeling approach that we described in [10]. In this chapter, we had to make assumptions on the power consumption and timing behavior within each state of the accelerometer, based on a combination of data sheets, scope measurements with a high-frequency oscilloscope and Powerscale measurements.

The idea behind the modeling method shown there is to express the inner states of the utilized hardware using *priced timed automata*. Each state has a power consumption associated with it. The question is, for how long will the hardware reside in each state? As this is dependent on the interaction of the components and the implementation of the driver—here given by two *interrupt service routines* (ISR)—we need to know exactly at what time either of the ISRs is triggered by the acceleration sensor and for how long they perform.

The system in this example behaves like that: After initialization of the system, the CPU goes to sleep mode and the acceleration sensor starts to measure its acceleration in three spatial directions. It does this at a frequency of 400 Hz. After each cycle, three 16-bit values, one for each spatial axis, are ready to be retrieved. The sensor will now inform the CPU by triggering an interrupt, thus waking up the CPU. Now, a sequence of six byte transfer is started to read the sensor values over a *Serial Peripheral Interface* (SPI). Each transfer is initiated by the CPU, which goes to sleep mode while the actual transfer is active. The transfer ends with an interrupt, waking up the CPU to process received data and initializing the next transfer. Figure 7 illustrates this.

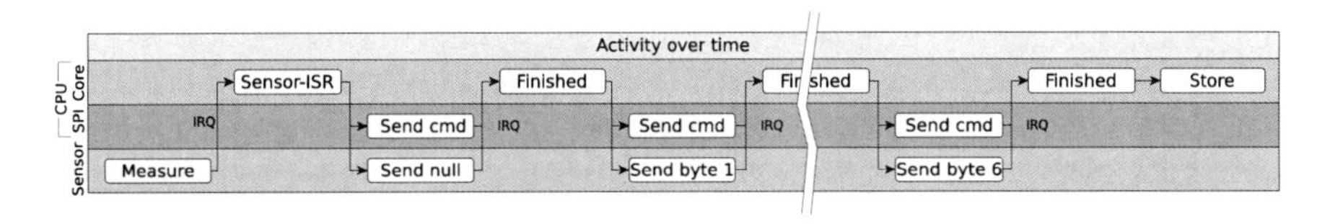

Fig. 7 Activity diagram of an accelerometer measurement

The last ISR call is used for post-processing the data and storing them into a fixed memory location where an application can access it. Then the CPU finally goes to sleep and waits for the next round.

Figure 6b shows the results of the measurement. The peaks (from left to right) reflect the behavior of the system: Three energy peaks for measuring the axes, followed by a phase of CPU activity in the first ISR. Afterwards a cycle of 6 double-peaks, where the smaller peaks shows the energy needed by the SPI transfer and the bigger peak showing the ISR. The last peak takes more time, as the collected data is written into memory.

The bottom curve shows the same measurement repeated using Powerscale. Although both devices use the same sample rate, the MIMOSA measurement shows a clearer signal shape.

7 Conclusion

In this chapter we described MIMOSA, a new approach to energy measurement in low-power systems. MIMOSA uses analog integration circuits to overcome the energy loss that is inherent to usual sampling devices, due to their inability to sense peaks between the samples. MIMOSA uses a 100 kHz sampling frequency for digitizing the output of the analog integrators. We could show that MIMOSA is able to sense peaks of at least 100 ns or frequencies of 10 MHz, respectively. We were also able to show that measurements in the 1 μJ region only show a signal noise of about 1 nJ (0,1 %). This is possible by utilizing a fast voltage-regulator to compensate any voltage drop at the shunt resistor, thus allowing to use high ohmic shunts, which have a high signal-to-noise ratio.

In a direct comparison and practical application, MIMOSA had to perform against Powerscale, an award-winning power measurement device that was designed and advertised for creating energy models of embedded system software routines. We were able to show that MIMOSA is able to outperform Powerscale in our experiments.

Acknowledgments This work was partly supported by the German Research Council (DFG) within the Collaborative Research Center SFB 876, project A4.

References

1. Nakutis, Z.: Embedded systems power consumption measurement methods overview. MATAVIMAI **2**, 29–35 (2009)
2. Hergenröder, A., Furthmüller, J.: On energy measurement methods in wireless networks. In: IEEE International Conference on Communications (ICC), pp. 6268–6272. IEEE (2012)
3. Jiang, X., Dutta, P., Culler, D., Stoica, I.: Micro power meter for energy monitoring of wireless sensor networks at scale. In: Proceedings of the 6th International Conference on Information Processing in Sensor Networks. IPSN '07, pp. 186–195. New York, NY, USA, ACM (2007)

4. Konstantakos, V., Kosmatopoulos, K., Nikolaidis, S., Laopoulos, T.: Measurement of power consumption in digital systems. IEEE Trans. Instrum. Meas. **55**, 1662–1670 (2006)
5. Analog Devices: Single-phase multifunction metering ic with di/dt sensor interface (ADE7753-C Data Sheet)
6. Maxim Integrated: Smart battery monitor (DS2438 Data Sheet)
7. Microchip Technology Inc: Energy-metering ics with active (real) power pulse output (MCP3905/06 Data Sheet)
8. Hitex Development Tools GmbH: Energy optimization: Powerscale. http://www.hitex.com/index.php?id=powerscale (2013). Accessed 17 June 2013
9. Agilent Technologies: N1911a/n1912a p-series power meters and n1921a/n1922a wideband power sensors (N1911A/N1912A/N1921A/N1922A Data Sheet)
10. Buschhoff, M., Günter, C., Spinczyk, O.: A unified approach for online and offline estimation of sensor platform energy consumption. In: 8th International Wireless Communications and Mobile Computing Conference (IWCMC), pp. 1154–1158 (2012)

A Remotely Programmable Modular Testbed for Backscatter Sensor Network Research

Eleftherios Kampianakis, John Kimionis, Konstantinos Tountas and Aggelos Bletsas

Abstract The necessity of backscatter sensor networks (BSNs) has recently emerged due to the need for large-scale, ultra low-cost, ultra low-power, wireless sensing. Development of such networks requires tools for rapid prototyping and evaluation of key-enabling BSN technologies. Although tools for testing wireless sensor networks (WSNs) have been widely developed over the last few years in the form of *testbeds*, almost no significant testbed examples exist for BSNs. Throughout this work, a set of hardware, firmware and software components have been designed and implemented, creating a BSN research testbed. The latter employs a modular architecture and enables rapid prototyping of critical components for low-cost, large-scale BSNs. Testbed components enable microwave, detection, coding and multiple access research, tailored for backscatter radio and networking. The testbed offers dynamic reconfiguration through implementation of remote, over the air programming (OTAP), that reduced programming time per node by two orders of magnitude. An overview of the testbed is given, and its modular tools are described in terms of functionality and importance for BSN research.

E. Kampianakis (✉) · J. Kimionis · K. Tountas · A. Bletsas
ECE Department,Technical University of Crete, Chania, Greece
e-mail: ekabianakis@isc.tuc.gr; kampianakis@gmail.com

J. Kimionis
e-mail: ikimionis@isc.tuc.gr

K. Tountas
e-mail: ktountas@isc.tuc.gr

A. Bletsas
e-mail: aggelos@telecom.tuc.gr

K. Langendoen et al. (eds.), *Real-World Wireless Sensor Networks*,
Lecture Notes in Electrical Engineering 281, DOI: 10.1007/978-3-319-03071-5_17,

1 Introduction

Technologies such as wireless sensor networks (WSNs) [1] and backscatter sensor networks (BSNs) [2, 3] lead towards large-scale, low-cost, wireless connectivity. Development tools for WSNs are widely developed, with typical examples being demonstrated in [4]. Testbed architecture includes wireless nodes under test, connected to interface boards acting as *monitors*. A gateway using a high level network interface (e.g Ethernet, 802.11) communicates with the interface boards.

On the other hand, limited research tools exist for BSN research and development. One example towards that direction is the work in [5], where a set of custom microwave devices for monitoring performance of radio frequency identification (RFID) antennas is presented. However, the setup is confined in measuring only microwave and antenna parameters.

This chapter describes the hardware and software components for the development of a BSN testbed. A prototype node with a single microcontroller unit (MCU) is developed that accommodates both a backscatter and a high level radio interface. The backscatter radio interface is implemented with a single RF transistor controlled by the MCU. The transistor acts as an antenna load switch, thus achieving backscatter modulation when an incident wave is reflected by the antenna [6]. For the high level control link, a 2.4 Ghz radio module is utilized.

The hybrid node developed (Fig. 1-left) acts both as the DUT, which is a semi-passive backscatter *tag* similar to the one presented in [7], and as the unit for remote programming, control and debugging functions. The DUT and monitor functions are constructed in software such that they are completely independent to one another.

The constructed testbed is shown in Fig. 1-right and its architecture is shown in Fig. 2. Remotely controlled Carrier emitters at the UHF band (868 MHz) illuminate the hybrid testbed nodes; the latter scatter back modulated information using a single transistor front-end. A software defined radio (SDR), tuned at the UHF band, acts as the tag receiver (reader). The SDR is connected to a host PC, where all signal processing takes place in software. The PC also hosts the testbed gateway, which transmits control packets and executable code to the hybrid nodes and the carrier emitters,

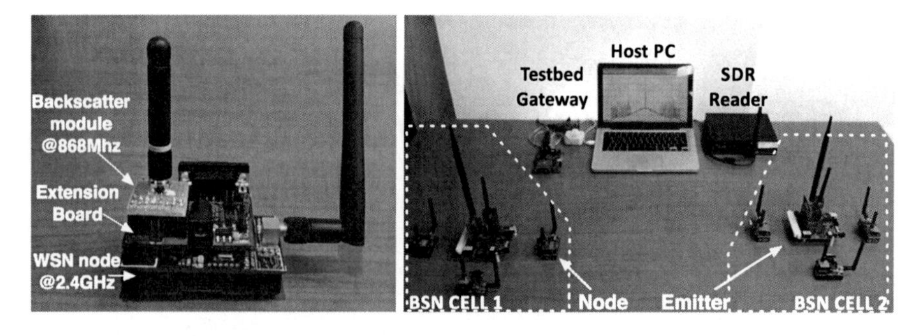

Fig. 1 *Left* Hybrid testbed node. *Right* Experimentation testbed

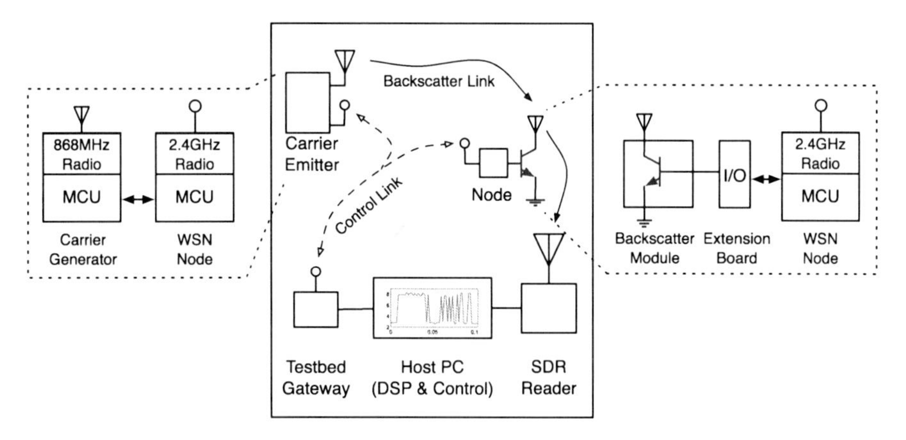

Fig. 2 *Center* Testbed overview. *Left* Carrier emitter. *Right* Hybrid node architectures

over a reliable 2.4 GHz link. It also receives debugging messages and feedback from the monitoring nodes. Instead of monostatic architecture, commonly used in RFID systems, bistatic backscatter radio is utilized (i.e emitter dislocated from receiver). Bistatic architecture allows extended communication ranges thus rendering broad area BSNs [8] viable.

This is the first effort towards the construction of a complete BSN testbed that is flexible, remotely programmable, and BSN research-oriented. Research topics such as detection schemes, resource allocation techniques, networking algorithms, as well as RF hardware (e.g switching transistors) and antennas where examined end evaluated with the utilization of this work.

The chapter is organized as follows: Section 2 describes the benefits of utilizing the BSN testbed. Section 3 gives the description of the hardware and software components required for the testbed implementation. Section 4 provides a set of research-oriented experimentation scenarios for testbed evaluation. Finally, Sect. 5 offers the conclusion of this work.

2 Benefits/Features of a BSN Testbed

The utilization of a BSN testbed offers engineers *mobility–portability*, *long range remote programmability*, *debugging and network monitoring*, high level of software and hardware *flexibility*, *dynamic reconfigurability*, with relatively *low cost* and most importantly, *reduced experimentation time*.

Portability is a critical feature for real-life, outdoor network testbeds. Figure 3 shows a testbed deployment outdoors, where environmental sensing applications are evaluated. Six hybrid nodes are place around a battery-operated carrier emitter and up to 17 m away from the reader. The SDR reader the host PC and a spectrum analyzer

Fig. 3 Remotely programmable testbed and hybrid node, outdoor deployment

(for debugging purposes) are placed in fixed position due to the power supply needed for the SDR. However , the latter could be modified for completely portable operation (e.g. see work in [9]).

Over the air programming (OTAP) is utilized in this work, as in many WSN testbeds. It allows wireless programming of network nodes with limited access, and accelerates the overall software development process. Particularly, with wired programming, for an outdoor, six node network such as the one in Fig. 3, a total time of 12 min (720 s including mobility) is required for programming the network. With OTAP utilization, programming time is reduced to 12 s (i.e two orders of magnitude reduction).

All components of the testbed are low-cost commodity solutions, or custom-built, in-house fabricated hardware. Particularly, backscatter radio hardware (i.e RF transistor, antenna) can be replaced and various sensors and actuators can be easily installed. Carrier emitters are fully configurable through software in terms of frequency and output power. Applications, from simple LED blinking to backscatter communication schemes are written in simple C language. Finally the total cost of the testbed depicted in Fig. 1-right is approximately 2,300 Euros, with the PC and SDR being the most expensive, while hybrid nodes cost an order of magnitude less.

3 Testbed Implementation

Hardware The core testbed module is a hybrid node.

Having the classic WSN testbed in mind a WSN node was utilized to act as the monitoring device and as the DUT. A module equipped with a backscatter RF front-end is connected to the interface board of the WSN node [10]. The node exchanges control data with the gateway using the embedded 2.4 Ghz RF module and backscatter communication is achieved with the backscatter module.

The backscatter module is equipped with a RF transistor and an SMA antenna connector. The transistor's base is driven by an MCU pin, while the other two are connected to the antenna terminals. When the MCU drives the base pin on high level

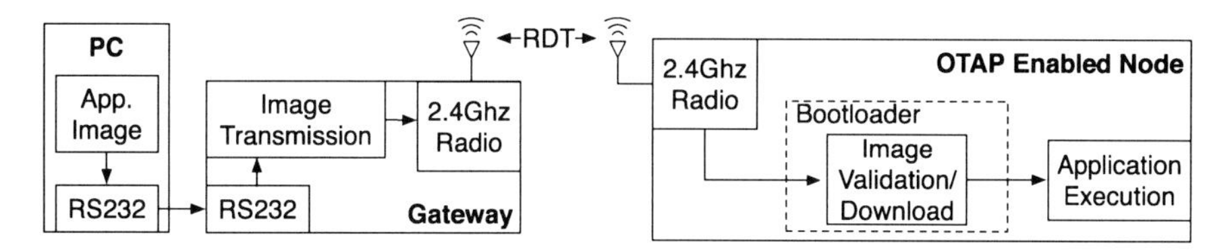

Fig. 4 Over the air programming block diagram

or low level, the transistor acts as a short or an open circuit, respectively. This allows switching between two antenna reflection coefficients, therefore enabling backscatter modulation [6, 11]. Figure 1-left depicts the implemented node and Fig. 2 (right) depicts node architecture.

To enable backscatter communication, low-cost, battery-operated RF carrier emitters are utilized. These devices, in the form of monolithic MCU-radio, are programmable signal generators. These modules are connected with the WSN nodes via the reliable 2.4 GHz link and as result, there is full control over the whole network. Several of the devices may exist on a test field, promoting experimentation with bistatic/multistatic backscatter links. These are less-explored backscatter architectures, with emerging research interest [8, 12, 13].

For the reception of backscatter signals, a commodity software defined radio (SDR) is used, while the processing takes place on a host PC, using MATLAB. This offers the flexibility to study communication schemes in depth. Since full control over the physical layer is required, no commercial "black box" devices are utilized.

Software The major software component of this work implements remote testbed programmability. The testbed's software system architecture is depicted in Fig. 4 and mainly consists of three parts: the bootloader, the gateway firmware, and the application. The application may be any type of code that can be executed by the node platform. Initially, the user application image file is transferred to the gateway from the host PC through an RS232 interface. The gateway firmware is responsible to reliably transmit the image file to all the network nodes within its range. Mechanisms for compressing image data in order to reduce the overall programming time, as well as reliable data transmission (RDT) protocols have been employed. The bootloader installed in each node (1) handles the wireless code reception via the 2.4 GHz interface, (2) validates the code, and (3) begins the program execution. The fact that the bootloader is the sole responsible module for programming the nodes renders its fail-safe operation critical. Thus, functions such as checksum image file validation and watchdog timing have been developed, in order to address any undesired situation. Finally, the bootloader is invoked by the application when a new program needs to be downloaded.

The above procedure, accelerates development, since the average time for wireless programming per node is about 2 s. Methods and devices that require physical contact are time consuming (more than 30 s per node) and messy.

4 Research with a BSN Testbed: Scenarios and Applications

High level modularity of the testbed's software and hardware components allows various experimentation possibilities. With the setup of Fig. 5, a set of research application scenarios have been implemented and evaluated on the proposed testbed. Six nodes have been used, however, the number of BSN nodes can be extended to several hundred.

RF/Microwave Research for BSNs Experimentation with various front-ends (i.e antennas and antenna loads) is necessary for the evaluation of tag design which greatly affects communication link performance [8]. Commodity or prototype antennas [14] can be quickly installed, and backscatter link budgets can be tested [12]. Figure 6 depicts the received power spectrum for a half-wavelength and a quarter-wavelength antenna, respectively, with the testbed node placed at a fixed location.

Physical Layer Communication Examination of communication schemes, receivers' performance and communication tradeoffs is feasible. On-off-keying (OOK) and frequency-shift-keying (FSK) modulations along with the corresponding receivers were developed and evaluated. Remote experimentation was conducted avoiding channel fading while full control of the testbed was maintained. Noise and received

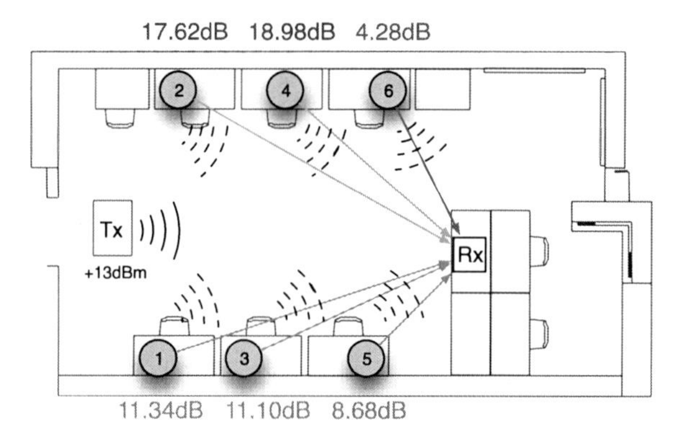

Fig. 5 Measured SNR values per backscatter node in lab deployment

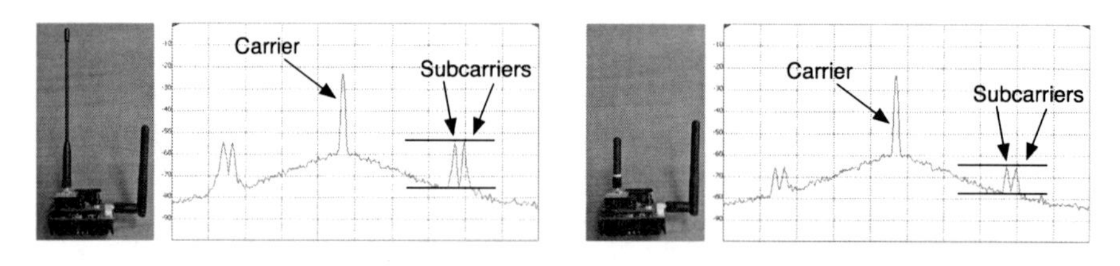

Fig. 6 Received power spectrum of a half-wavelength and a quarter-wavelength antenna with fixed tag-reader distance. The switching (i.e *subcarrier*) frequencies are clearly visible. The quarter-wavelength antenna backscattered signal has fairly smaller amplitude than the one backscattered from the half-wavelength antenna

carrier power were estimated at the reader, followed by a per-tag estimation of the backscattered signals' SNR. The estimated SNR values for carrier power of +13 dBm can be seen in Fig. 5, employing OOK and a nonlinear receiver. Tags were switched on and off sequentially to ensure that only one was transmitting at a time.

Finally various idiosyncrasies of bistatic backscatter links were examined such as the carrier frequency offset between carrier emitters and the reader and the "range-versus-rate" communication tradeoffs. Figure 7 depicts the baseband received spectrum of a carrier emitter along with a tag performing FSK modulation, with 100 and 125 kHz subcarriers. One can observer the increased noise floor around the carrier, which affects negatively the tags' SNR, since its subcarriers reside inside the clutter frequency region. Communication was practically impossible, with a measured BER value of 35 %. Followingly the tag was reprogrammed to operate in different subcarriers (250, 300 kHz), away from the clutter. The resulting BER value was 3.5 %.

Cross Layer Experimentation There are problems that require experimentation with all the layers of the testbed. Such is the study of the *capture effect*, when two nodes utilize the same communication channel simultaneously and collide. However under certain conditions, demodulation can be achieved, if the weaker signal is regarded as noise. Theoretical research on the field exists [15], but real-system study is difficult due to synchronization and control issues. Figure 8-left shows the forced collision of two tags, making the demodulation impossible. After the OTAP (Fig. 8-right), two *distant* tags are forced to collide. The tag closer to the reader, the one with higher SNR value, is demodulated, while the second tag can be seen as noise on the top of the strong tag's waveform.

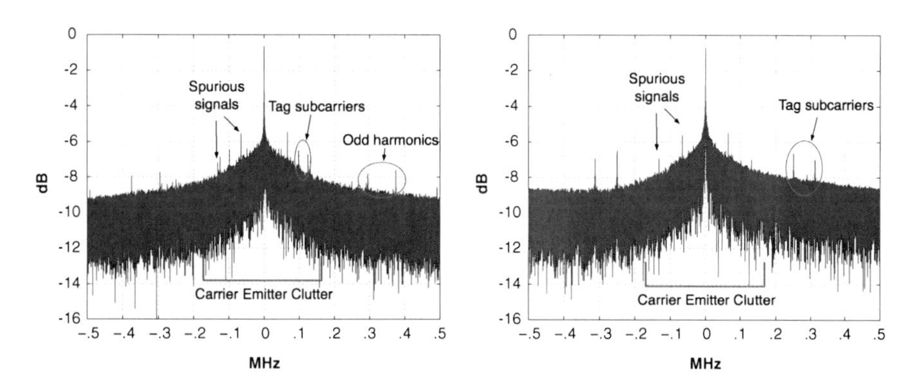

Fig. 7 *Left* tag operating inside RF clutter (BER 35 %). *Right* tag operating away from RF clutter (BER 3.5 %)

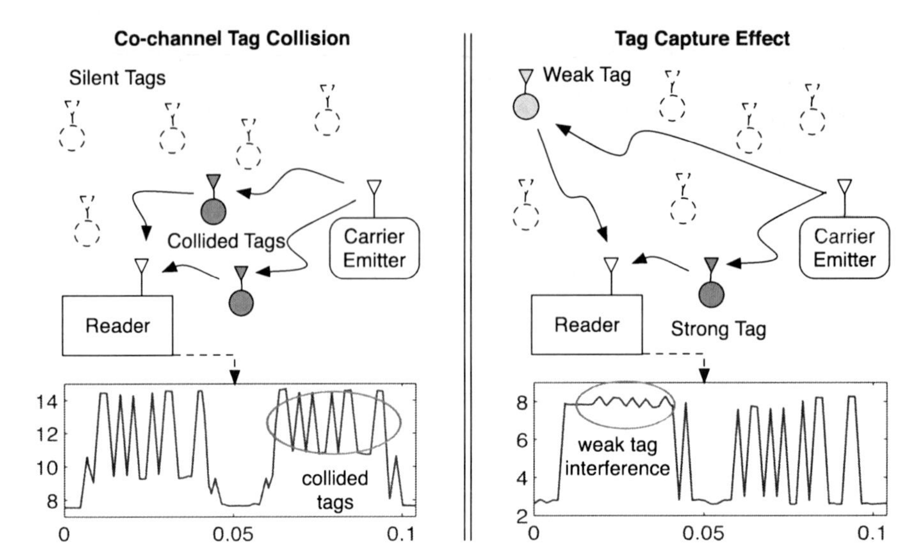

Fig. 8 Capture effect experimentation. *Left* Tags collide by transmitting at the same time. *Right* Tags are re-programmed to examine capture effect (strong tag is decoded successfully, even at the presence of a weak interfering tag). The figure depicts real signals captured by the SDR receiver

5 Conclusion

This chapter describes in detail the design and implementation of a low-cost backscatter sensor network testbed that merges WSN and BSN technologies. Hardware/software flexibility and over the air testbed programmability, facilitate simple debugging and monitoring of critical backscatter communication parameters. Finally, real world, indoor or outdoor deployments demonstrated testbed mobility and portability.

References

1. Akyildiz, I., Su, W., Sankarasubramaniam, Y., Cayirci, E.: Wireless sensor networks: a survey. Comput. Netw. **38**(4), 393–422 (2002)
2. Vannucci, G., Bletsas, A., Leigh, D.: Implementing backscatter radio for wireless sensor networks. In: Proceedings IEEE Personal, Indoor and Mobile Radio Communications, 1–5 Sept 2007
3. Vannucci, G., Bletsas, A., Leigh, D.: A software-defined radio system for backscatter sensor networks. IEEE Trans. Wireless Commun. **7**(6), 2170–2179 (2008)
4. Werner-Allen, G., Swieskowski, P., Welsh, M.: Motelab: a wireless sensor network testbed. In: Fourth International Symposium on Information Processing in Sensor Networks (IPSN), pp. 483–488 April 2005
5. Griffin, J., Durgin, G., Haldi, A., Kippelen, B.: How to construct a test bed for RFID antenna measurements. In: IEEE Antennas and Propagation Society International Symposium, pp. 457–460 Jul 2006

6. Dobkin, D.M.: The RF in RFID: Passive UHF RFID in Practice. Elsevier, Amsterdam (2008)
7. Thomas, S., Reynolds, M.: QAM backscatter for passive UHF RFID tags. In: IEEE International Conference on RFID, pp. 210–214 Apr 2010
8. Kimionis, J., Bletsas, A., Sahalos, J.N.: Bistatic backscatter radio for tag read-range extension. In: IEEE International Conference on RFID-Technologies and Applications (RFID-TA), Nice, France, Nov 2012
9. Dickens, M., Dunn, B., Laneman, J.N.: Design and implementation of a portable software radio. IEEE Commun. Mag. **46**(8), 58–66 (2008)
10. Bletsas, A., Vlachaki, A., Kampianakis, E., Sklivanitis, G., Kimionis, J., Tountas, K., Asteris, M., Markopoulos, P.: Building a low-cost digital garden as a telecom lab exercise. IEEE Pervasive Comput. **12**(1), 48–57 (2013)
11. Bletsas, A., Dimitriou, A.G., Sahalos, J.N.: Improving backscatter radio tag efficiency. IEEE Trans. Microw. Theory Tech. **58**(6), 1502–1509 (2010)
12. Griffin, J.D., Durgin, G.D.: Complete link budgets for backscatter-radio and rfid systems. IEEE Antennas Propag. Mag. **51**(2), 11–25 (2009)
13. Bletsas, A., Kimionis, J., Dimitriou, A.G., Karystinos, G.N.: Single-antenna coherent detection of collided FM0 RFID signals. IEEE Trans. Commun. **60**(3), 756–766 (2012)
14. Kruesi, C.M., Vyas, R.J., Tentzeris, M.M.: Design and development of a novel 3D cubic antenna for wireless sensor networks (WSN) and RFID applications. IEEE Trans. Antennas Propag. **57**(10), 3293–3299 (2009)
15. Lai, Y.C., Hsiao, L.Y.: General binary tree protocol for coping with the capture effect in RFID tag identification. IEEE Commun. Lett. **14**(3), 208–210 (2010)

Part IV
Networking

A Scalable Redundant TDMA Protocol for High-Density WSNs Inside an Aircraft

Johannes Blanckenstein, Javier Garcia-Jimenez, Jirka Klaue and Holger Karl

Abstract We present the results of a measurement campaign conducted with a wireless sensor network (WSN) deployment inside an aircraft. A robust and scalable TDMA protocol for mission-critical applications was developed, which exploits spatial diversity provided by redundant access points. The WSN, consisting of 500 sensor nodes organized in three cells with two redundant access points per cell, was installed in an Airbus A330–300. The link quality and the packet error rate with and without the redundant access points was evaluated. It was found that the packet error rate could be decreased more than four times by using the spatial diversity introduced by the dual access point approach.

1 Introduction

The applications of wireless sensor networks in aircraft are manifold [1], e.g. cabin status monitoring, maintenance support, structural health monitoring and location-aware mobile devices. Thus, the requirements are diverse in terms of data rate, delay, and loss tolerance. All applications should be served by the same wireless infrastructure and communication protocol. The protocol must be able to support many devices with a high density, since there are potentially several thousands of wireless sensors in a single aircraft. The available bandwidth is scarce, especially considering low-power wireless standards like IEEE802.15.4, where only 225 kbit/s maximum theoretical data rate is available for a single-hop single-channel network, while with multi-hop this data rate is at least divided by the number of hops [2].

J. Blanckenstein (✉)· J. Garcia-Jimenez · J. Klaue
EADS Innovation Works, Munich, Germany
e-mail: johannes.blanckenstein@eads.net

H. Karl
University Paderborn, Paderborn, Germany
e-mail: holger.karl@upb.de

K. Langendoen et al. (eds.), *Real-World Wireless Sensor Networks*,
Lecture Notes in Electrical Engineering 281, DOI: 10.1007/978-3-319-03071-5_18,

Additionally, using the standard slotted CSMA/CA mechanism the maximum data rate is limited to around 35 kbit/s [3] or even 12 kbit/s [4] with around 50 active wireless nodes. It is our goal to support a high number of nodes (thousands) within a fairly small space while providing defined data rates, delay bounds, and loss rates for different applications. For this purpose a robust MAC protocol for mission-critical applications is needed. In [5], MAC protocols are classified according to the network performance parameters delay and reliability. Only a few protocols fall in the appropriate class "delay-intolerant and loss-intolerant", namely QoS-MAC [6], MMSPEED [7], Burst [8] and GinMAC [9]. All reviewed delay and loss-intolerant protocols are based on TDMA. The survey [5] also states that all reviewed protocols except WirelessHART [10] are based on academic or theoretical studies and might not be suitable for real deployments. Delay bounds and packet delivery guarantees are of course always based on the assumption that the wireless channel is not too bad all the time. Based on channel models derived from aircraft measurement campaigns [11–15] we derived a maximum distance between sender and receiver of 10 m. For an efficient use of the channel capacity we abstain from multi-hop and ACK/NACK-based retransmissions as well. Since bad channel conditions cannot be avoided even over a quite short distance, robustness is introduced by redundant access points (AP). The additional diversity is basically for free since most packets are broadcast uplink (from node to APs). In addition to the spatial diversity from distributed access points, each AP has two antennas. Our MAC-protocol is based on a planned TDMA scheme which makes use of the scalable number of redundant APs. It is described in Sect. 2.

In order to test our design goal of reliable communication with strict delay bounds while using the channel capacity efficiently, we deployed 500 wireless sensor nodes in an Airbus A330–300. The measurement campaign as well as the developed wireless sensor network platform is described in Sect. 3. In Sect. 4 we show the results of the performance evaluation in terms of received signal strength and packet error rates.

2 Scalable Redundant TDMA

In dense WSNs with high traffic, contention is likely. Therefore, the medium access control (MAC) mechanism is a critical factor in the design of the system. TDMA has the advantage that channel capacity utilization, network delays, and power consumption can be determined in advance.

We propose a TDMA MAC protocol which fulfills our requirements in dense WSNs. These are achieved, in part, through the concept of AP redundancy instead of retransmissions. An efficient synchronized single-hop scheduling, based on a pure TDMA protocol, is established in a centralized and scalable way. Up to eight redundant APs are supported per cell. The protocol uses a periodic time-slotted superframe structure that only consists of guaranteed time slots forming the contention-free periods for the beacons and logical links. We present the protocol as a directed graph G(B, L, C, D, U, I) in Fig. 1, where B denotes the set of beacons, L the set of logical links, C the number of APs, D the number of downlink slots, U the number of uplink slots,

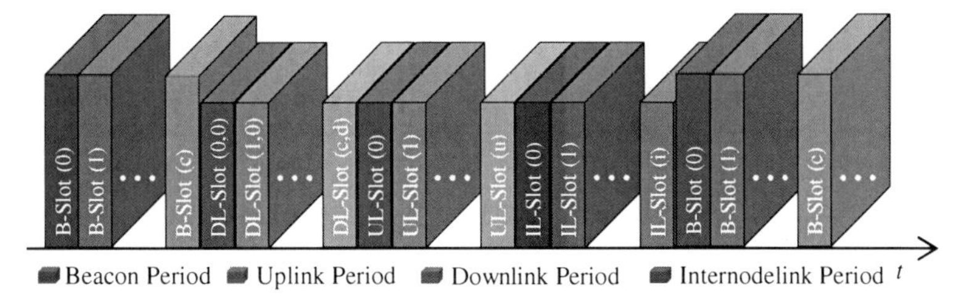

Fig. 1 TDMA protocol structure

and I the number of internode-link slots. Let B(c) denote the beacon from AP c, DL-Slot(c, d) denote the downlink slot d from AP c to the nodes, UL-Slot(u) the uplink slot u from node to the APs and IL-Slot(i) the internode-link i from node to nodes. The slot duration and the set of slots forming each logical link period (D, U, I) can be defined and optimized, both statically or dynamically, according to the specific network requirements, such as throughput, delay, and node allocation.

For synchronizing with the TDMA network, a low duty cycle listening mode based on a CCA algorithm has been implemented with the protocol on the sensor node. The operating AP with the lowest slot allocation will provide the initial synchronization to the rest of the devices. Thus, each AP has to be in the communication range of at least one lower-slot-allocation peer. Beacons are broadcast by each AP. These contain all the necessary information to synchronize with the superframe and schedule node activity within it. For reliable synchronization and data delivery, redundancy is introduced in the system by sending multiple copies of beacons and downlink messages from spatially distributed APs and by broadcasting to these APs in the uplink. From an energy perspective this method does not lead to a significant increase of power consumption. Nodes will only listen until the first copy of the data packet was correctly received. Afterwards, they sleep during the rest of the redundant slots.

3 Measurement Campaign

An experimental measurement campaign has been performed with an implementation of this TDMA protocol. The aim was to characterize the wireless environment in an aircraft and to confirm the improvements that the protocol provides.

3.1 Setup

The network was organized in three cells with two redundant APs per cell. The positions of the APs were selected, on the one hand, to provide maximum coverage and, on the other hand, to resemble likely installation positions. The sensor nodes

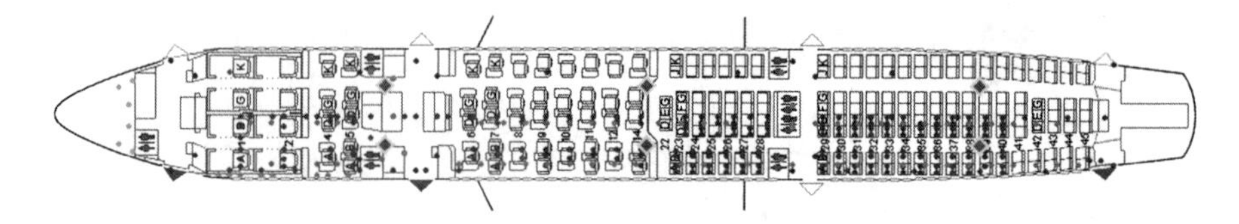

Fig. 2 Setup of access points and wireless sensor nodes in the aircraft

were installed at the positions where the potential applications require them, for instance at the seats, life vest compartments, light units, lavatories, kitchens, galleys and cargo compartments. Each dot in Fig. 2 represents one node, the colors symbolize different applications. However, for the purpose of this chapter, all of them are treated uniformly. The nodes were mostly placed on the left side of the aircraft due to its symmetry. Equipping all seats on the right side as well would only lead to mirrored conditions regarding the opposed APs. The six red squares represent the APs. Two APs close together form a cell and work in the same radio-frequency channel. Assigning a different channel for each cell enables interference-free, parallel operation of the cells.

The sensor nodes are assigned to cells according to their distance from the APs with a maximum distance of 10 m. The aircraft is approximately 60 m long so that three cells with a radius of 10 m cover the whole aircraft. A maximum number of 170 nodes per cell was defined for the campaign, so that the superframe of a cell consists of two beacons (one from each AP), two downlink command slots and 170 uplink slots (one for each sensor). With a slot length of 5 ms a superframe is 870 ms long. A sensor needs to receive at least one beacon and one downlink command and to transmit one packet. On average a packet is 33 bytes or 1 ms long which means that the duty cycle is $3/870 = 0.36\,\%$ most of the time. In the worst case, which is when the first beacon is missed and also the first command is missed, the node needs to listen for four whole slots leading to a duty cycle of $(4 \cdot 5 + 1)/870 = 2.4\,\%$.

Figure 3 shows the periodic message flow of the measurement protocol developed as a simple method to assess the quality indicators for a complete network scenario. A server triggers the APs via IP/UDP multicast datagrams to broadcast the *measurement requests* according to the defined TDMA protocol. Nodes respond by sending their measured data in their slot and the APs forward the answers back to the server.

Each node measured temperature, battery voltage and received signal strength (RSSI) for the *measurement requests* received from each of the two access points. The APs measured the RSSI for each received packet. That means that 4 RSSI values $(AP_0 \rightarrow Node, AP_1 \rightarrow Node, Node \rightarrow AP_0, Node \rightarrow AP_1)$ were measured for each node per superframe. In total around 38 million transmissions were recorded. The packet error rates for the up- and downlink were calculated for each AP alone and for both combined. In the downlink, "combined" means a node received a request from either AP; and in the uplink, either AP received the packet from the node. Based on that the packet error rate (PER) was calculated for the four links separately and for the combined links. Since the access points are designed with two

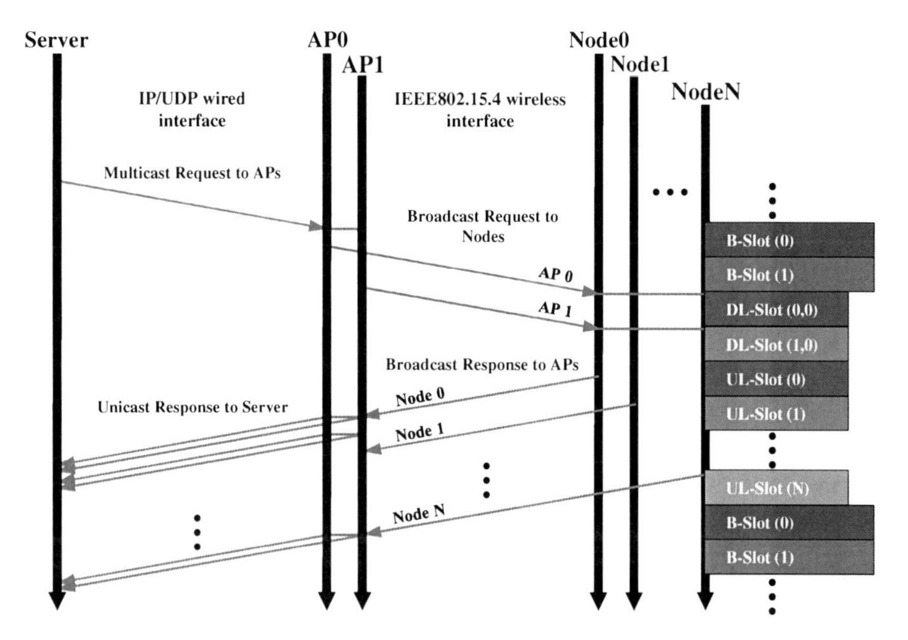

Fig. 3 Message flow-chart

antennas, measurements were done with either antenna and additionally with the antenna auto-select mode of the transceiver. In the following section the PERs are compared for the different antennas as well as for the redundant APs in order to evaluate the impact of antenna selection versus that of spatial diversity.

3.2 Low-Power Wireless Sensor Node Platform

As a testbed platform, sensor nodes were specifically developed for the aircraft applications. Table 1 summarizes its main features.

In a TDMA scheme, the duty cycle has the major effect on the energy consumption of the sensor node. To observe this impact, an analysis was carried out on increasing the superframe duration by adding more uplink slots, while keeping the configuration defined in the previous section (two redundant APs and one downlink slot). In this way, the duty cycle is reduced since sensor nodes are only active during beacon and downlink slots and during the transmission in their allocated uplink slot. The configured RF parameters for the sensor node are -101 dBm RX sensitivity and $+4$ dBm TX output power. The results of these measurements are presented in Fig. 4.

The graph reveals that with increasing superframe length (lower duty cycle), the averaged power consumption approaches 22 μA asymptotically. This value is the minimum required current by the board when all components are in deep sleep

Table 1 Wireless sensor node data sheet

Processor	ARM CM3 EFM32G210F128
SRAM	16 KB
Flash	128 KB
System	FreeRTOS
Transceiver	IEEE802.15.4 2.4 GHz AT86RF233
Antenna	Fractus Compact Reach Xtend 1 dBi
Built-in sensors	Humidity and Temperature Sensor SHT21
Power source	Rechargeable Lithium Polymer 350 mA @ 3.7 V
Weight	3.6 g PCB
	21.4 g incl. battery and housing
Dimensions	50 × 21 × 0.4 mm PCB
	55 × 24 × 17 mm housing
Temperature range	−40 to +85 °C

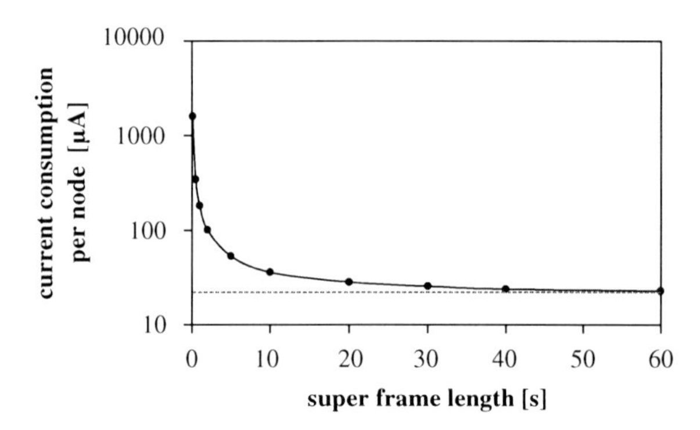

Fig. 4 Current consumption regarding superframe length

mode. This consumption is dominated by the power management chip necessary for the LiPo battery.

3.3 Wireless Access Point

To provide a complete WSN solution, an access point was also developed as IP gateway and coordinator for the wireless protocol. Among its most important features, showed in Table 2, we point out the support of antenna diversity, to reduce RF effects of multipath propagation and fading, and PoE supply, to facilitate the deployment of the network.

Table 2 Wireless access point data sheet

Processor	ARM926EJ
	ARM CM3 EFM32GG330F1024
SRAM	4 MB + 128 KB
FLASH	8 MB + 1 MB
System	Embedded Linux + FreeRTOS
Transceiver	2.4 GHz AT86RF233
Antenna	Dual diversity RP-SMA Connector
Power source	PoE (Power over Ethernet)
Weight	47 g PCB
	178 g incl. housing and antennas
Dimensions	76 × 63 × 1 mm PCB
	94 × 134 × 46 mm housing
Temperature range	−40 to +85 °C

4 Results

First, the quality of the wireless channel is evaluated with respect to RSSI values and PER. Second, the impact of dual access points and different antenna selection modes is evaluated.

4.1 Base Measurements

The RSSI values obtained by the transceiver during the measurement campaign are in the range of −91 to −10 dBm with a resolution of 1 dBm and an accuracy of 3 dBm. In Fig. 5, the average RSSI value per node over the distance between the access point and the corresponding node is shown, together with four aircraft channel measurements from literature [12–15] and the free-space model. As can be seen, the RSSI values form a dense cloud around the channel measurements from literature. But in contrast to them, a trend line for the RSSI values would follow an even steeper slope than suggested by the models. This is due to shadowing caused by real application positions, e.g. underneath the seats. Additionally, some nodes in places like the cargo compartment or the electronic bay have even lower RSSI values than the majority of the nodes. These nodes form the scarce cloud below −70 dBm. Based on the measured distribution of the RSSI values we reason that a cell size of 10 m still leads to acceptable RSSI values. With a bigger cell radius the received signal strength will decrease to a level where it will have significant impact on the packet error rate for the current system.

Figure 6 shows the difference of averaged RSSI values per node; in Fig. 6a the mean RSSI values of AP 1 are subtracted from AP 0 and in Fig. 6b the mean RSSI values of the downlink are subtracted from the uplink. The difference of the values

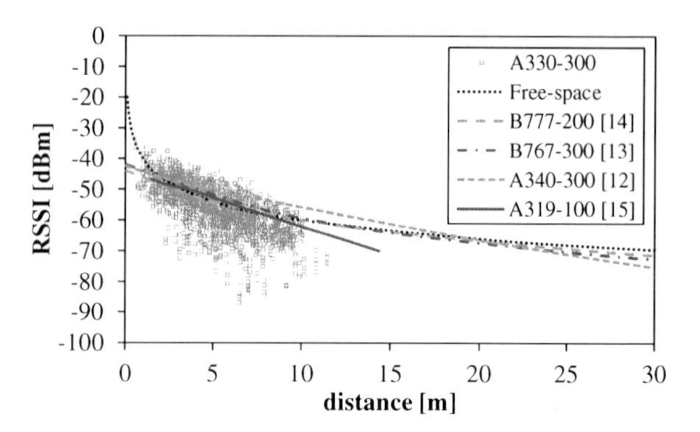

Fig. 5 Average RSSI per node compared to channel measurements from literature

can indicate if the channel is biased. In Fig. 6a, the RSSI values deviate around a common mean (≈ -2.1 dB); the received signal strength for AP 1 is slightly higher. As outlined in Sect. 3.1 the nodes are not equally distributed throughout the aircraft and therefore the distance to Access Point 1 is a little bit shorter on average; hence the higher RSSI values. Nevertheless, there are lots of nodes for which the signal strength is higher at AP 0. This emphasizes the benefit of the multi-AP approach to utilize spatial diversity. In Fig. 6b the difference of the RSSI values of uplink−downlink is shown. The deviation of the difference is smaller than in Fig. 6a and in the region of the accuracy of the RSSI values. We expected this deviation, but with a mean of 0 instead of -2.6 dB. The RSSI values for the downlink were always higher than for the uplink. Most probably it is caused by non-reciprocal components in the access points, but this has to be validated with further measurements. Figure 7 shows the packet error rate for all nodes in the uplink case. Figure 7a shows the uplink packet error rate with an abscissa scaling up to 1. As can be seen, for most of the nodes the PER is quite low with the exception of some outlier nodes. The moderate outliers occur when the link quality is poor due to "challenging" placements of the nodes. The outliers with high values result from nodes which lost connectivity to the APs for a long time. In Fig. 7b the abscissa is scaled up to a PER of 0.01 and outliers above a threshold of 0.01 are removed to improve the visualization. As can be seen, the packet error rate varies over the nodes, but the combined PER (green area) is always lower than the PER when just using AP 0 or AP 1. With two access points per cell in the uplink case the combined PER is more than four times lower than compared to the PER when only using either AP 0 or AP 1. Because of the high PER of the outliers the mean is not a well suited measure for the tendency of the PER. When comparing the mean value for $a = 1.1 \cdot 10^{-2}$ (with outliers) with $b = 4.9 \cdot 10^{-4}$ (without outliers) the discrepancy is obvious. Instead, the median for both cases equals $2.5 \cdot 10^{-4}$ and is therefore used in the remainder of the chapter as the measure for comparing PERs.

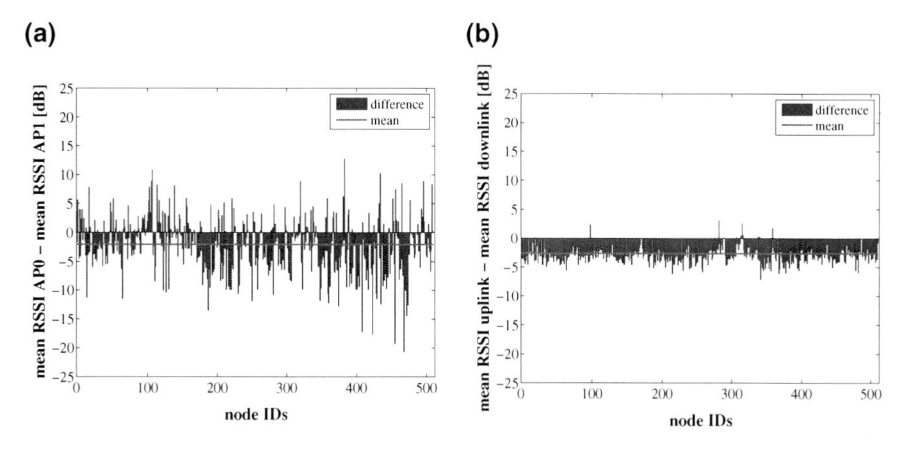

Fig. 6 **a** Difference of the mean RSSI values of AP 0 and 1, and **b** Difference of the mean RSSI values of uplink and downlink

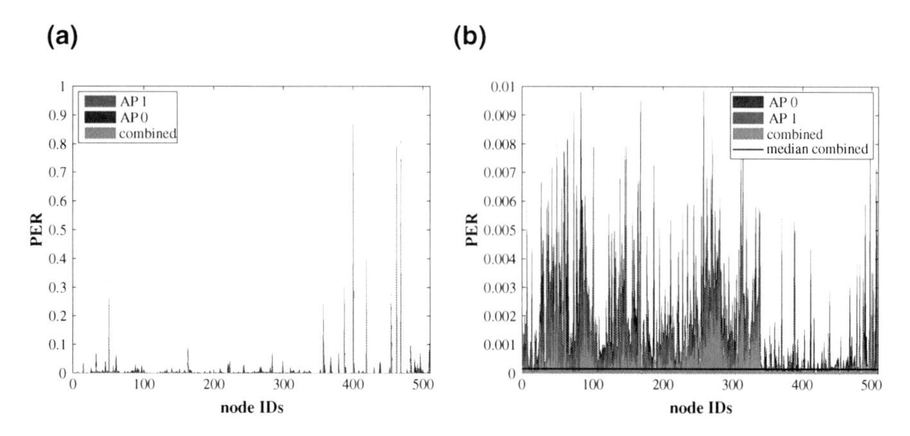

Fig. 7 **a** PER for the uplink, and **b** PER for the uplink with all outliers (above 0.01) removed

4.2 Impact of Dual Access Points

Two APs are responsible for each cell. Downlink packets—beacons included—are sent by both APs. Uplink packets are broadcast and because of the cell size can be received by both APs. Therefore, in both directions spatial diversity and in the downlink also temporal diversity is exploited.

To gain as much diversity as possible, the best case would be when the links to AP 0 and AP 1 were statistically independent. Then, the combined PER would equal the product of the individual PERs of the APs, ($PER_C = PER_{AP0} \cdot PER_{AP1}$). To evaluate the dependency of those links, we calculated the cross-correlation of the RSSI values and determined the *Pearson product-moment correlation coefficient* per node. A high correlation coefficient means that both APs measure similar RSSI

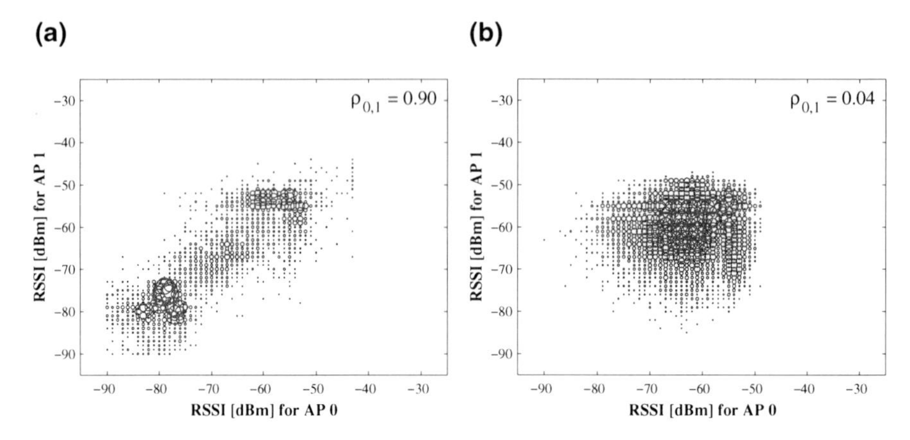

Fig. 8 Time-aligned RSSI values for AP 0 and 1 in the downlink; **a** RSSI values for node 165 (high correlation coefficient), and **b** RSSI values for node 45 (low correlation coefficient). The diameter of the circles is proportional to the number of measurements

values, while a low coefficient means that their measurements are independent from each other. Figure 8 visualizes this dependency with the help of two examples. The links for the depicted node in Fig. 8a are statistically dependent; both APs measure similar RSSI values at each time, which therefore form a cloud around the bisecting line. Statistically independent links are depicted in Fig. 8b. For this node, no correlation can be observed for the measured RSSI values. The values do not follow the bisecting line, but they seem to be equally distributed. The links to both APs are slightly correlated, but this correlation is still low enough to exploit diversity. The mean correlation coefficient over all nodes in the downlink equals 0.30 with a standard deviation of 0.42, while for the uplink the mean equals 0.21 with a standard deviation of 0.31. The high deviation indicates a different potential gain for each node by the multi-AP approach; nevertheless, for most of the nodes the impact will be high. Figure 9 shows the median of the PERs for the downlink, the uplink, and for the round-trip communication. The PER for AP 1 is in all three cases slightly higher than the PER for AP 0. In the downlink 18 %, in the uplink 9 %, and for the round-trip 17 %, even though AP 1 received a 2.1 dB stronger signal. As can be seen in Fig. 5, the RSSI values are in the region of -40 to -70 dBm, therefore even with a 2.1 dB weaker signal the signal-to-noise ratio is still high and will have no big influence on the PER. The same is valid when comparing the downlink with the uplink. The 2.6 dB stronger signal in the downlink also has no significant influence on the PER.The combined PER will be lowest, if both links are statistically independent. With the given PERs for AP 0 and AP 1, this lower bound is in the order of 10^{-6}. The measured combined PER was in the order of 10^{-4}. Nevertheless, in all three cases the combined PER is much lower than with only one AP, which clearly shows the benefit of the multi-AP approach when introducing spatial diversity. The round-trip PER is of course higher than only for the down- or uplink. Nevertheless, the combined PER

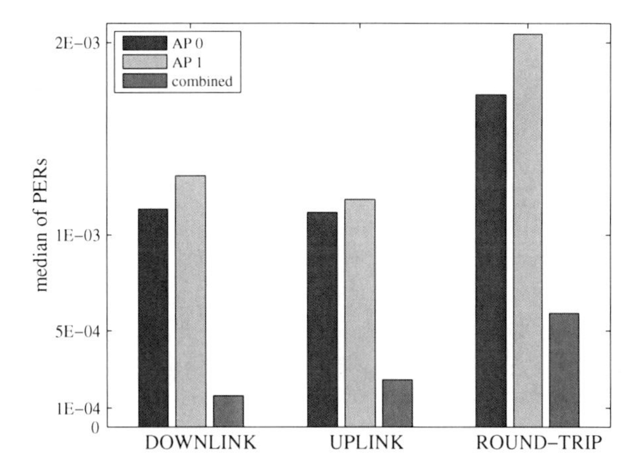

Fig. 9 Median of PER for AP 0, 1, and combined

in the round-trip case is still lower than the PERs for AP 0 or 1 in down- and uplink individually.

4.3 *Impact of Antenna Selection*

As explained in Sect. 3.3 the access point is equipped with two antennas, where only one of them can be selected at a time. There are three different antenna selection (AS) modes available: AS 1 for always using the first antenna, AS 2 for always using the second antenna, and AS 3, the antenna auto-select mode, where the transceiver chooses the antenna. For more detail on auto-select mode, see Sect. 11.4 in the data sheet of the transceiver Atmel AT86RF233. We made sure that always the same AS mode was selected for all access points. To evaluate the impact of the AS modes in Fig. 10a the median of the packet error rates for different antenna selection modes are depicted. The minimal resolution of the PERs is a consequence of the limited measurement time. This minimal resolution is depicted as the black bar within each PER bar. As can be seen, for mode AS 3 the measured PER is identical to the resolution of these measurements and therefore can only be used as the upper bound. The PER for the antenna selection mode AS 2 is always slightly higher than for AS 1. In Fig. 10b the ratio of the packet error rates for up- and downlink are shown. As can be seen, for all combinations the ratio is nearly one, which means that each chosen antenna selection mode has the same impact on the PER even though the downlink has a 2.6 dB higher RSSI than the uplink. It becomes obvious that the multi-AP approach has a much higher impact on the PER compared to the impact of different antenna selection modes. Because of the limited time for the measurements the impact of different antenna selection modes could not be investigated with the required resolution. This will be done in future measurement campaigns.

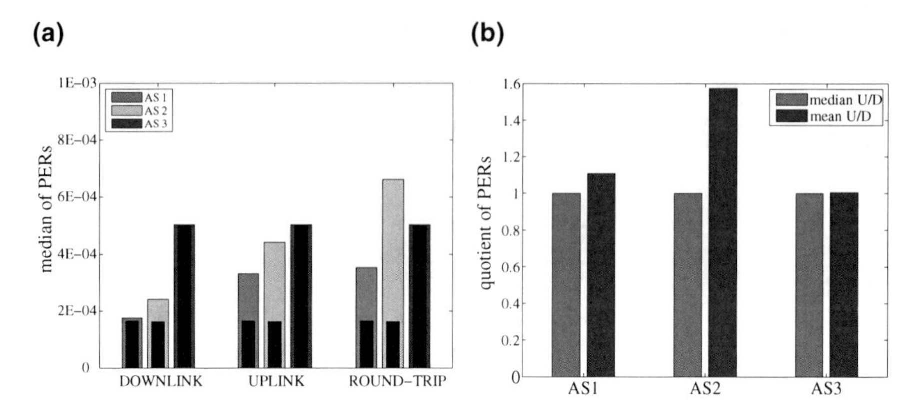

Fig. 10 **a** Median PER for AS modes, and **b** Ratio of PER for up- and downlink

5 Conclusions

A wireless sensor network with 500 sensor nodes was successfully deployed in an Airbus A330–300. The designed TDMA protocol abstains from multihop and from ARQ-based retransmissions in order to provide guaranteed maximum delay while maintaining a high channel utilization. Robustness is achieved by spatial redundancy provided by multiple APs. The measured environment is even more challenging than suggested by channel models from literature, which becomes obvious when comparing the measured distribution of RSSI values with the models. The reason for that is the deployment of the nodes at the application positions, e.g. in the life vest compartments under the seats. Two APs per cell were used to introduce spatial diversity. This diversity reduced the PER significantly. The access points were equipped with two antennas, but the measurement results did not show significant impact on the PER when selecting one or the other.

A new measurement campaign will be conducted to evaluate the decrease of PER when using more than two access points per cell and to find out when increasing spatial diversity stops improving PER. Additionally, the extended measurement campaign will be used to gather more data. This is necessary to increase the resolution of the PER and to be able to investigate the impact of antenna selection more thoroughly.

References

1. Lewandowski, A., Michaelis, S., Wietfeld, C., Klaue, J., Kubisch, M.: In-cabin localization solution for optimizing manufacturing and maintenance processes for civil aircrafts. In: Position Location and Navigation Symposium (PLANS), 2012 IEEE/ION, pp. 1257–1264 (2012)
2. Österlind, F., Dunkels, A.: Approaching the maximum 802.15.4 multi-hop throughput. In: Proceedings of the Fifth ACM Workshop on Embedded Networked Sensors (HotEmNets 2008), June 2008
3. Petrova, M., Riihijarvi, J., Mahonen, P., Labella, S.: Performance study of ieee 802.15.4 using measurements and simulations. In: IEEE Wireless Communications and Networking Conference, 2006, WCNC 2006, vol. 1, pp. 487–492 (2006)
4. Lee, T.-J., Lee, H.R., Chung, M.Y.: MAC throughput limit analysis of slotted CSMA/CA in IEEE 802.15.4 WPAN. IEEE Commun. Lett. **10**(7), 561–563 (2006)
5. Suriyachai, P., Roedig, U., Scott, A.: A survey of MAC protocols for mission-critical applications in wireless sensor networks. IEEE Commun. Surv. Tutorials **14**(2), 240–264 (2012)
6. Suriyachai, P., Roedig, U., Scott, A.: Implementation of a MAC protocol for QoS support in wireless sensor networks. In: IEEE International Conference on Pervasive Computing and Communications, 2009, PerCom 2009, pp. 1–6 (2009)
7. Felemban, E., Lee, C.-G., Ekici, E.: MMSPEED: multipath multi-SPEED protocol for QoS guarantee of reliability and timeliness in wireless sensor networks. IEEE Trans. Mobile Comput. **5**, 738–754 (2006)
8. Munir, S., Lin, S., Hoque, E., Nirjon, S.M.S., Stankovic, J.A., Whitehouse, K.: Addressing burstiness for reliable communication and latency bound generation in wireless sensor networks. In: Proceedings of the 9th ACM/IEEE International Conference on Information Processing in Sensor Networks, IPSN '10, pp. 303–314. New York, USA (2010)
9. Suriyachai, P., Brown, J., Roedig, U.: Time-critical data delivery in wireless sensor networks. In: Proceedings of the 6th IEEE international conference on Distributed Computing in Sensor Systems, DCOSS'10, pp. 216–229. Springer, Heidelberg (2010)
10. Chen, D., Nixon, M., Mok, A.: WirelessHART: Real-Time Mesh Network for Industrial Automation, 1st edn. Springer, NewYork (2010)
11. Kaouris, A., Zaras, M., Revithi, M., Moraitis, N., Constantinou, P.: Propagation measurements inside a B737 aircraft for in-cabin wireless networks. In: IEEE Vehicular Technology Conference 2008, VTC Spring 2008, pp. 2932–2936 (2008)
12. Moraitis, N., Constantinou, P.: Radio channel measurements and characterization inside aircrafts for in-cabin wireless networks. In: IEEE 68th Vehicular Technology Conference, 2008. VTC 2008-Fall, pp. 1–5 (2008)
13. Hankins, G., Vahala, L., Beggs, J.H.: Propagation prediction inside a B767 in the 2.4 GHz and 5 GHz radio bands. In: IEEE Antennas and Propagation Society International Symposium, 2005, vol. 1A, pp. 791–794 (2005)
14. Hankins, G., Vahala, L., Beggs, J.H.: 802.11ab propagation prediction inside a B777. In: IEEE/ACES International Conference on Wireless Communications and Applied Computational Electromagnetics 2005, pp. 837–840 (2005)
15. D'Errico, R., Rudant, L.: UHF radio channel characterization for wireless sensor networks within an aircraft. In: Proceedings of the 5th European Conference on Antennas and Propagation (EUCAP), pp. 115–119 (2011)

Do We Really Need a Priori Link Quality Estimation?

Vasilis Vasilopoulos, Daniele Puccinelli and Marco Zúñiga

Abstract Traditionally, link quality estimation (LQE) has been viewed as an a priori step in sensor network routing protocols because it filters out unreliable links before data transmission. Recent results, however, show that protocols can perform well without a priori LQE. Because getting rid of LQE seems rather counter-intuitive, the aim of this work is to look deeper into the behavior of LQE-free protocols. Our results, based on one of the state-of-the-art LQE-free protocols, show two interesting insights. First, LQE-free protocols manage to choose links that are slightly better than the ones obtained with a priori LQE methods. Second, in traditional protocols, the effort needed to identify good links accounts, on average, for roughly half of the energy consumption of nodes, depending on the nodes' active period and on the inter-packet interval. By eliminating this overhead, LQE-free protocols can save a significant amount of energy compared to standard approaches.

1 Introduction

Traditional collection protocols in sensor networks, such as CTP [1] and Arbutus [2], employ a two-step approach to build their routing structure. First, nodes perform link quality estimation (LQE) to get a sense of all the possible usable links. Then, each node forms a routing path on top of these usable links (according to a given metric). Performing link quality estimation makes sense because unreliable links are commonplace in sensor networks (due to the use of low-cost radios) and it is important to filter them out to increase reliability.

In the last couple of years, however, a new generation of collection protocols [3–6] have challenged this traditional view, and they have achieved better energy efficiency

V. Vasilopoulos · M. Zúñiga (✉)
Delft University of Technology, Delft, The Netherlands
e-mail: m.a.zunigazamalloa@tudelft.nl

D. Puccinelli
University of Applied Sciences and Arts of Southern Switzerland, Manno, Switzerland

K. Langendoen et al. (eds.), *Real-World Wireless Sensor Networks*,
Lecture Notes in Electrical Engineering 281, DOI: 10.1007/978-3-319-03071-5_19,

than CTP-like approaches. While all these new protocols have their own distinctive features, they share a common characteristic: they do not perform a priori link quality estimation. This observation motivated us to look deeper into the (in)significance of link quality estimation for sensor networks.

The most important task of sensor networks is the ability to deliver data. Link quality and path length play a central role in this respect: the better the link and the shorter the path, the higher the delivery rate. But, if no a priori link quality estimation is performed, several important questions arise. Without LQE, how good (or bad) are the selected links? What is the impact on the path length? The new generation of collection protocols reports delivery rates that are comparable or higher than CTP; therefore, they must be utilizing good links. Does this imply that a priori LQE is not necessary?

Furthermore, since link quality estimation typically comes with a sizable control traffic overhead and, therefore, a potentially significant energy footprint, it is important to quantify the exact energy demands of LQE methods (since the elimination of these extra energy demands may be the core reason why the new generation of protocols outperform the older one).

In this chapter we report some preliminary results that are part of a wider research effort aiming at understanding LQE-free protocols. We selected the Broadcast-Free Collection Protocol (BFC [6]) as an initial step to delve into the questions posed above. Our empirical evaluation provides some interesting insights:

- LQE-free protocols choose links that are as good as, or even slightly better than those chosen by a priori LQE methods. While our results are based on BFC, we conjecture that this behavior also holds for other protocols (at least in some scenarios).
- We analyze the energy budget of CTP, dividing it between the parts used for control traffic (mainly LQE) and data transmissions. Our results show that the control traffic part uses between 40 and 60 % of the energy resources of nodes (depending on the nodes' active period and on the inter-packet interval). Quantifying this overhead is central to understand the (in)efficiency of methods aiming at identifying good links.

2 Preliminaries

In recent years, several collection protocols have been proposed that do not rely on LQE as heavily as in previous research efforts, which treated it as an indispensable building block. Before describing the related work in detail, it is important to highlight that this new generation of LQE-free protocols provides the same delivery rate as CTP [1] and most of these new protocols do so by consuming less energy.

The summary of the state-of-the-art is captured in Table 1. For the duty cycle we do not provide the actual numbers, but rather a relative metric (since most of the corresponding chapters used different evaluation scenarios). For example, the value

Table 1 Performance of recent collection protocols

	BCP	LWB	ORW	BFC	CTP
Relative duty cycle (%)	>5	0.34	0.4	0.53	1
Delivery rate (%)	>97	>99	>99	>99	>99

of 0.4 for ORW indicates that the duty cycle of this protocol is on average 60 % less than in CTP (as reported by the authors).

2.1 Related Work

The Backpressure Collection Protocol (BCP) [3] employs a data-driven approach and forwards packets to the neighbor with the lowest queue level. It is a notable example of structure-free routing, but it is not a great fit for duty-cycled sensor networks because it requires a significant level of offered load to be effective, and this increases the energy consumption of nodes (see Table 1).

The Low-Power Wireless Bus (LWB) [5] describes a way to perform data dissemination without utilizing LQE. LWB effectively turns a multi-hop low-power wireless network into a shared bus, where all nodes are potential receivers of all data. LWB leverages constructive interference to perform ultra-cheap and ultra-light network wide floods. The main limitation of this protocol with respect to CTP is that it is centralized; one node in the network is responsible for synchronizing the fast floods of all the other nodes.

The aforementioned protocols are great examples of new radical ways to disseminate and collect data without a priori LQE. Their mechanisms are, nevertheless, fundamentally different from CTP. BCP requires an artificially high offered load to bootstrap the system and LWB is centralized. To have a good (and fair) insight into the benefits of LQE-free approaches, we believe that it is better to first use protocols that, in spirit, are closer to CTP. Below we describe two of such protocols.

In Opportunistic Routing in Wireless Sensor Networks (ORW) [4], the sink first performs floods to form a gradient centered around itself. The nodes then use opportunistic unicast transmissions to forward data to the first neighbor that wakes up and that provides progress towards the sink (i.e. a node with a higher gradient). ORW does not rely on a priori LQE methods except for one instance, when nodes can not find a parent. The authors describe their approach as a "coarse-grained flavor of LQE".

The Broadcast-Free Collection protocol (BFC) [6] completely avoids LQE while building a data collection tree. BFC relies on eavesdropping nodes that already have a path towards the sink. The protocol is bootstrapped by the sink neighbors, which upon transmitting their first data packets get direct acks from the sink. The eavesdropping process is repeated by nodes further down the tree until all nodes get a path. If a node is unable to eavesdrop on a potential parent, it remains disconnected until a potential parent appears in its vicinity.

Selecting the best candidate for comparison As mentioned in Sect. 1, this chapter describes our first effort aiming at understanding LQE-free protocols. As a starting point we required a protocol that resembles CTP as much as possible but that requires no a priori LQE. ORW and BFC are similar to CTP, but ORW still performs some level of LQE (albeit minimal). To avoid obtaining wrong insights, such as getting LQE help from ORW when a node cannot find a parent, we decided to use BFC as a stepping stone.

Novelty This study is the first to identify that many of the newly proposed collection protocols have a common characteristic: the lack of LQE. Considering that we use BFC as a stepping stone to understand LQE-free protocols, it is important to highlight how our work differs from the study presented by Puccinelli et al. [6]. In that work, the authors propose BFC, provide an analytical model for the duty cycle of different types of nodes and compare BFC with CTP with respect to energy, delivery rate and latency. Their evaluation results do not look into the routing structure (link quality and path length) resulting from the lack of LQE methods. Also, while the authors do provide a thorough evaluation of energy consumption, they do not compare it with a dissected energy budget from CTP, i.e. CTP control traffic separated from CTP data traffic. This dissection is important because, as it will be explained later (see Sect. 3.4), it allows us to compare LQE-free protocols with the "LQE-free" version of CTP (which is obtained if the period of the Trickle timer in CTP increases to infinity). What is more, the above-mentioned dissection allowed us to trace a bug in the BFC code that was penalizing BFC's energy performance substantially. Finally, while the authors do mention that BFC may result in unbalanced networks, we provide data that supports their hypothesis. Overall, our goal is to better understand the properties of LQE-free operations using BFC as a case study of LQE-free operation and CTP as a case study of LQE-based operation.

2.2 Evaluation Methodology

Our experimental evaluation is based on the Indriya testbed [7]. We selected this testbed because it offers reasonably challenging testbed conditions as well as a relatively large-scale network (138 active nodes as of spring 2013).

We used existing TinyOS implementations of CTP and BFC, and ran both protocols on top of BoX-MAC [8]. In all of our experiments nodes inject one data packet every 5 min (a reasonable interval value for several data collection applications). The sink was located at one edge of the network (node 1 in the top right corner) and it was always on (a typical choice since the sink is normally connected to a station with unlimited access to energy resources). Given that sensor network testbeds can have different connectivity patterns at different times [9], we always ran BFC and CTP back-to-back so that both protocols could operate under similar conditions. We used a transmit power of 0 dBm on 802.15.4 channel 26 in all experiments.

Even though we ran experiments over several LPL wake-up interval values, in this chapter we show the results for a wake-up interval of 1 s because it is in the optimal operating range of CTP—with respect to energy consumption—for the assumed offered load of 1 packet per node every 5 min. For all the other LPL intervals (>1 s), BFC performed even better than CTP. In line with most experimental studies in this area, we use the duty cycle as a proxy for energy consumption [10].

In the next section, we report our results based on the different roles that nodes play in the network. This is done because the sink neighbors, relays and leaves can have vastly different performances. The sink neighbors are the nodes that can reach the sink in one hop, relays are the nodes that are not sink neighbors but that need to forward packets for other nodes (besides their own packets), and leaves are the nodes that only have to send their own packets.

3 Experimental Study

Our evaluation has two overarching goals. First, to understand the impact of eliminating a priori LQE on the underlying routing structure (link quality and path length). Second, to quantify the impact of LQE methods on the network's energy consumption (duty cycle and load balance). Overall, our evaluation captures four messages:

- LQE-free protocols seem to select links that are as good as those chosen by LQE methods.
- The paths of LQE-free protocols tend to be longer than those found in LQE-based protocols.
- By eliminating LQE, protocols can save a substantial amount of energy. For LPL wake-up interval of 1 s and inter-packet interval of 5 min, this translates into a reduction of energy expenses between 40 and 60 %.
- Depending on the method, LQE-free protocols may lead to a better or worse load balance. In the case of BFC, the load balance and, therefore, the energy balance is worse.

3.1 Link Quality

To utilize a common metric, we define link quality as the average number of transmissions required to send successfully a packet from any node to its immediate parent. Figures 1 and 2 show the cumulative distribution function (*cdf*) for all child-parent links for relay and leaf nodes, respectively, and Table 2 reports the mean and standard deviation. We observe that performing a priori LQE does not help in selecting better links. In fact, BFC tends to select better links. CTP's links are roughly 9 % costlier than BFC's for relays and 5.5 % costlier for leaves on this occasion. Below we provide a conjecture as to why this happens. We do not show the results for the sink neighbors because all of these nodes had perfect links with CTP and BFC.

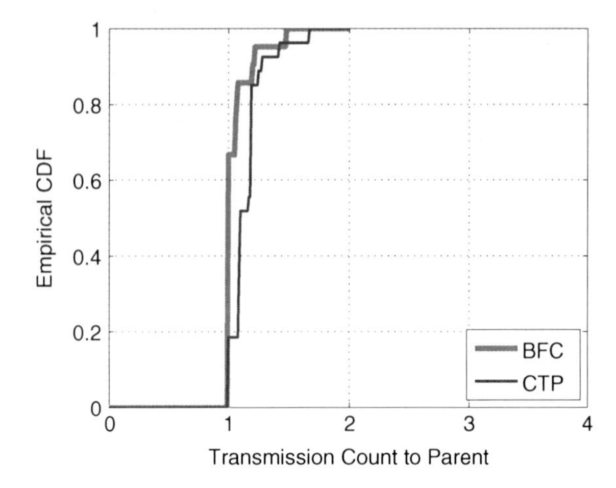

Fig. 1 CDF of the average number of transmissions to the selected parent for the relays

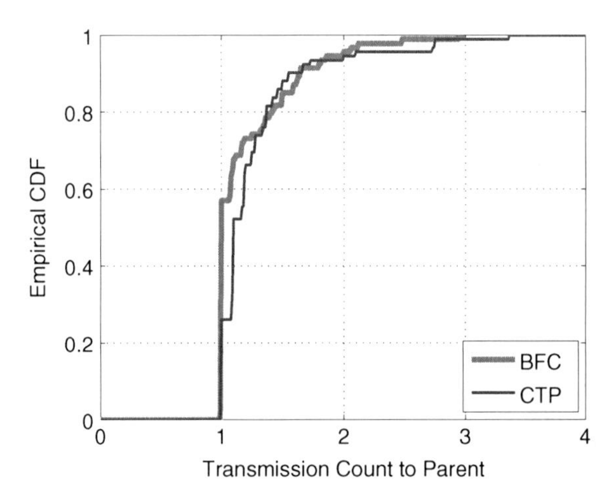

Fig. 2 CDF of the average number of transmissions to the selected parent for the leaves

Table 2 Link quality (# transmissions)

	Relays		Leaves	
	CTP	BFC	CTP	BFC
Average (μ)	1.15	1.05	1.27	1.20
Std dev (σ)	0.14	0.11	0.42	0.36

Why does a priori LQE not seem to matter? The key limitation of LQE methods is the presence of "intermittent" links. Based on the various LQE studies in the literature [11], let us assume the following simplified classification of sensor networks links: (i) good, when they rarely drop a packet (very high signal-to-noise ratio, way above the radio sensitivity), (ii) intermittent, when they have periods of good quality but also periods of bad quality (the signal-to-noise ratio oscillates around the radio sensitivity), (iii) unreliable, when most of the time their quality is below good (the signal-to-noise ratio is usually below the radio sensitivity). LQE methods are very good at discarding unreliable links, but not so good at discarding intermittent links (because there is a chance that the quality is measured during a good period). In fact, Alizai et al. identified this limitation as a weakness of CTP but from a different perspective [12]. That study claimed that CTP was too conservative. By discarding some of these intermittent links, CTP was not taking advantage of those links' good periods, which decreased its performance. In this study, we move in a different direction. We conjecture that LQE-free protocols overcome the lack of LQE methods by being even more conservative on the utilization of intermittent links.

In BFC, nodes have a pre-determined parent (as in CTP), but the network relies solely on good links. In BFC, a node is considered a parent if three consecutive data packets arrive from the (potential) child node. Notice that BFC does not use a single beacon to find a parent, BFC simply relies on sniffing and latching on nodes that already found a path towards the sink. Given that (i) in most data collection applications the period between consecutive data packets is of a few minutes, and that (ii) intermittent links are bursty in nature [11]; by validating a link every few minutes, the chances of picking an intermittent link are significantly reduced. BFC is conservative in the sense that, by default, its data path validation mechanisms discard unreliable and intermittent links. This conservative approach, letting BFC assume symmetric links, may explain why BFC selects slightly better links than CTP.

In ORW [4], nodes do not have a pre-determined parent; they transmit information opportunistically using a unicast paradigm. In this way, nodes do not assume that a link exists (which is what protocols with a priori LQE methods do); a node uses what is available at the moment. Given that link quality is known to be highly correlated in time, ORW usually ends up selecting either good links or intermittent links during good periods. ORW is conservative in the sense that unless a link is currently good, it assumes that it does not exist.

3.2 Path Length

In CTP, the a priori LQE phase not only allows the network to identify good links, but this preliminary information is also used to identify short paths. On the other hand, in LQE-free protocols nodes do not have this extra information, and hence, they may end up using longer routes. Figures 3 and 4 show the *cdf* of path lengths for relays and leaves, and Table 3 reports the mean and standard deviation. The absence of LQE clearly affects the depth of the data collection tree. Thanks to LQE, CTP's routes are roughly 8 % shorter for relays and 13 % shorter for leaves on this occasion. Even though we did not perform experiments with ORW, we expect this trend to be the same. Due to the opportunistic nature of ORW transmissions, packets can follow paths that deviate significantly from the shorter routes identified by CTP.

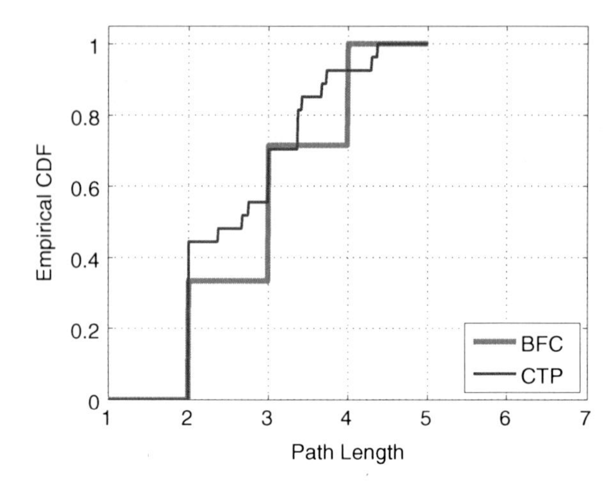

Fig. 3 CDF of the average number of hops to the sink for the relays

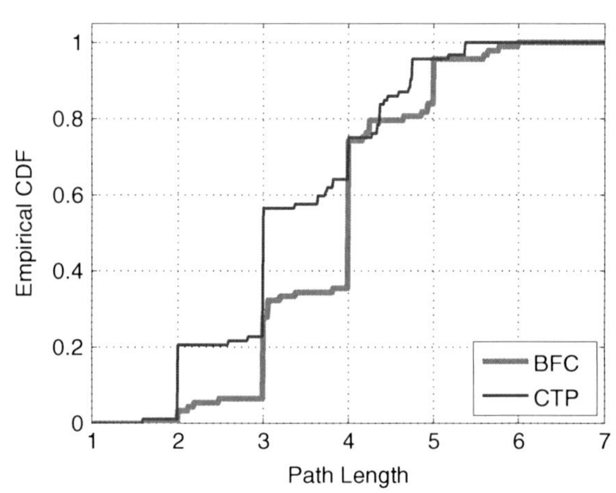

Fig. 4 CDF of the average number of hops to the sink for the leaves

Table 3 Path length (# hops)

	Relays		Leaves	
	CTP	BFC	CTP	BFC
Average (μ)	2.72	2.95	3.36	3.85
Std dev (σ)	0.77	0.80	0.99	0.87

3.3 Link Quality Versus Path Length

Link quality and path length are jointly responsible for the formation of the data gathering tree, i.e. building the routing structure. To provide a comprehensive view of the relation between link quality and path length, Figs. 5 and 6 show the routing cost (number of transmissions required to deliver a packet to the sink) versus the

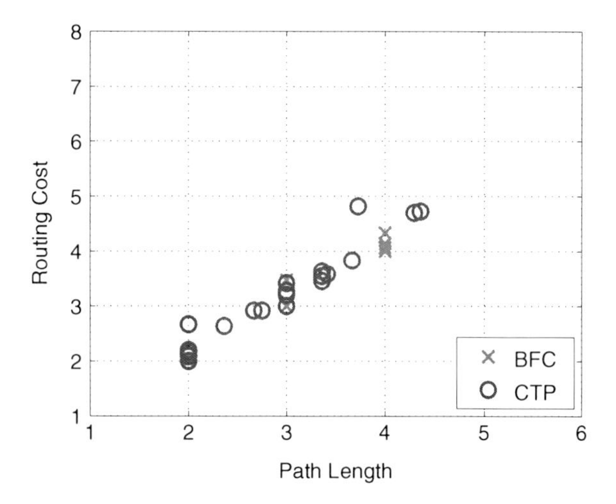

Fig. 5 Routing cost versus path length for the relays

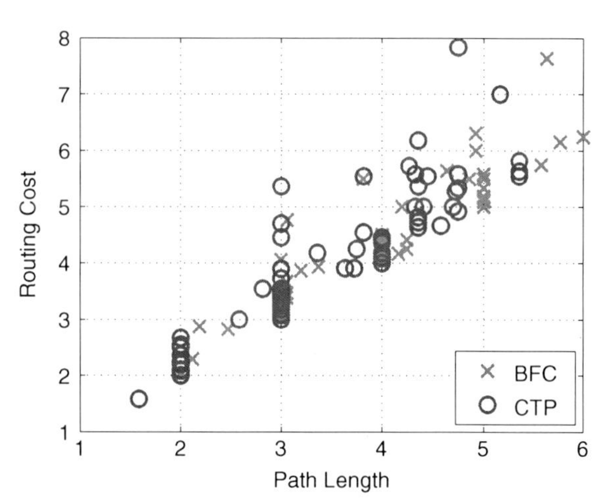

Fig. 6 Routing cost versus path length for the leaves

path length for relays and leaves. If all paths were error-free, the plot would show a line with slope 1. At first glance, we could state that LQE-methods are more efficient because they provide shorter paths with similar link qualities, that is, they require less transmissions to deliver the same amount of data. But as we will observe next, these savings are minimal compared to the energy expenses of CTP control traffic (which is required to find these good short routes in the first place).

3.4 Energy Efficiency

We have seen that without employing LQE underneath, BFC still opts for good links. Now, we will proceed to understand why the overall operation of BFC, and LQE-free

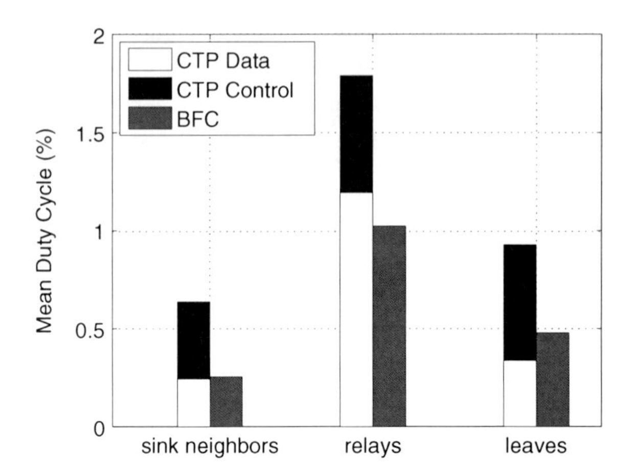

Fig. 7 Mean duty cycle for sink neighbors, relays and leaves

protocols in general, can be so energy-efficient. To this end, we changed slightly the CTP code to monitor the fractions of the duty cycle related to control traffic and to data traffic. We also found a bug in the BFC code that penalized its duty cycle (i.e. BFC's energy performance is better than the measurements reported in [6]). Figures 7 and 8 show the mean and median duty cycle for the three types of nodes. The median plot allows us to filter out connectivity outliers.

Figure 8 shows that the cost of CTP control traffic is similar for all node classes, in the order of 0.5 %. This occurs because CTP periodically sends probes to test the connectivity of nodes. It is important to highlight that these measurements were performed once CTP and BFC reached a stable state, that is, after the initial probing required to bootstrap the network (otherwise the duty cycle of CTP would be even higher). The important lesson from this result is that it gives us an "LQE-free" version of CTP (captured by the white bars). Assuming an ideal static environment, CTP would need to run the Trickle timer only once (at the beginning). Then, after the routes are formed, the LQE phase would not need to be run again because links would not change. This result allows us to benchmark an LQE-free protocol, such as BFC, with an ideal "LQE-free" version of CTP.

It is important to note that BFC has some other disadvantages besides the long latency mentioned in [6]. A priori LQE also allows a better responsiveness. CTP would react much faster to changes in the environment than BFC. For instance, if a malicious interfering node is first turned on and then off. CTP would be able to rapidly (re)identify the shorter paths "erased" by the interfering node. On the other hand, BFC, would not be able to recover (or at least not fast enough) because no periodic LQE measurements are done. As part of our future work, we plan to use the JamLab tool [13] to stress LQE-free protocols under different interference patterns.

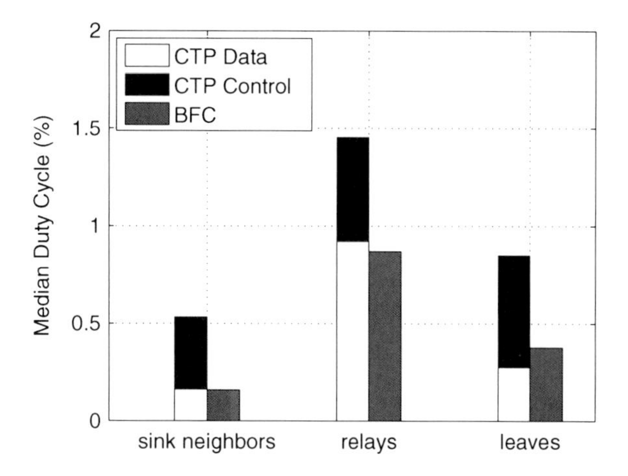

Fig. 8 Median duty cycle for sink neighbors, relays and leaves

3.5 Load Balancing

In [6], the authors mention that BFC seemed to have a worse load balance than CTP. We validate this hypothesis. Figures 9 and 10 illustrate the *cdf* of the forwarding load for the sink neighbors and relays. The forwarding load is defined as the ratio of the total number of forwarded packets (which includes locally generated and relayed packets) per locally generated packet. We do not present the *cdf* for the leaves because the vast majority of them had a forwarding load of 1. Strictly speaking, all the leaves should have a forwarding load of 1, but, in a real network, leaves may occasionally act as relays for short periods of time. We, therefore, use a threshold of 2 to filter out such nodes from the set of real relays.

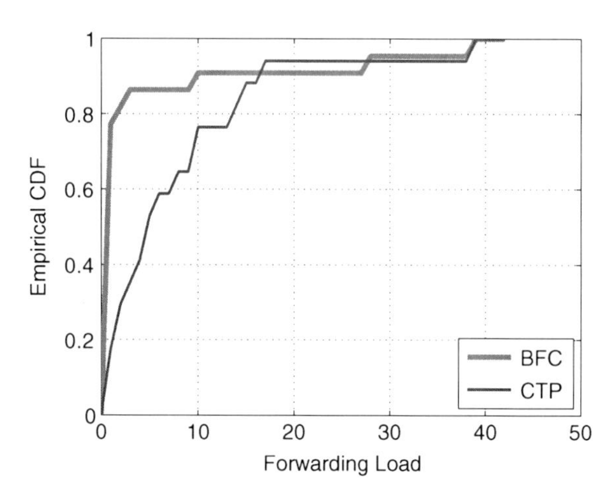

Fig. 9 CDF of the forwarding load for the sink neighbors

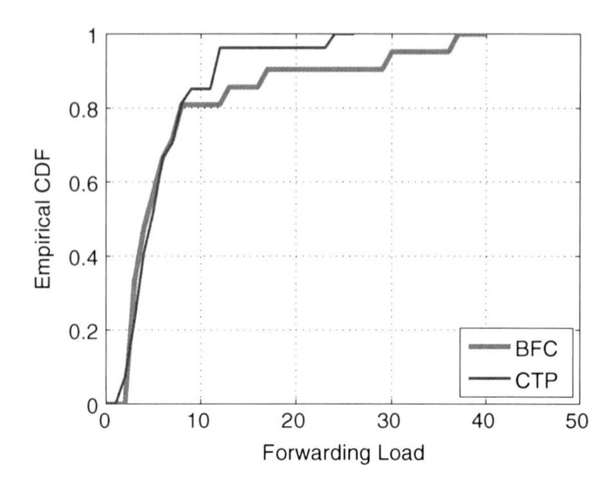

Fig. 10 CDF of the forwarding load for the relays

Table 4 Jain's fairness index

	# Sink neighbors	Index	# Relays	Index	# Leaves	Index
CTP	17	0.44	27	0.66	92	0.95
BFC	22	0.18	21	0.44	93	0.96

Recalling that neither CTP nor BFC employ an explicit load balancing scheme, we see that BFC performs clearly worse than CTP with respect to sink neighbors (see Fig. 9). A few sink neighbors in BFC are responsible for forwarding 80 % of the load. The load balancing is better for the relays (see Fig. 10). To quantify the unfair share of the forwarding load and consequently of the energy resources, we compute the Jain's fairness index for both CTP and BFC for all node classes in Table 4. A Jain's index of 1 denotes perfect balance (best-case value), while $1/n$ (where n denotes the number of nodes) indicates clear unfairness (worst-case value). If we consider all nodes, the Jain's fairness index is 0.29 for CTP and 0.18 for BFC.

The reason for the worse load balance in LQE-free approaches seems to be the same as for the longer paths (see Sect. 3.2). Without a priori LQE, a network not only looses the ability to identify good links but also to have a broader view of the underlying communication graph. This broader view could facilitate the identification of shorter paths and better load balancing.

3.6 Energy Efficiency Versus Load Balancing

To provide a complete view of the effect of LQE on the energy consumption of individual nodes, Figs. 11 and 12 show the duty cycle versus the forwarding load for the sink neighbors and relays. Leaves can be regarded as relays that are responsible

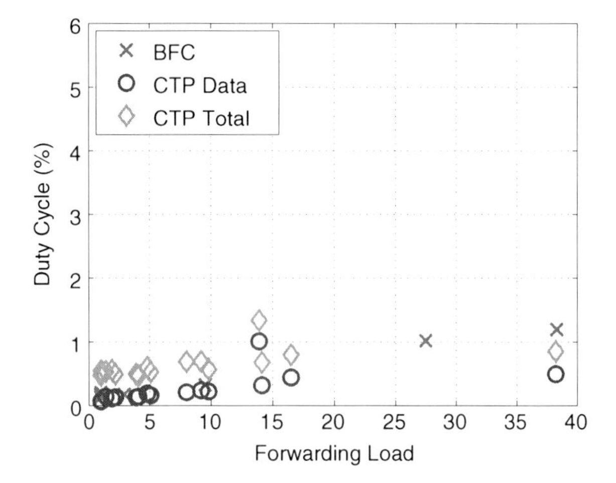

Fig. 11 Duty cycle versus forwarding load for the sink neighbors

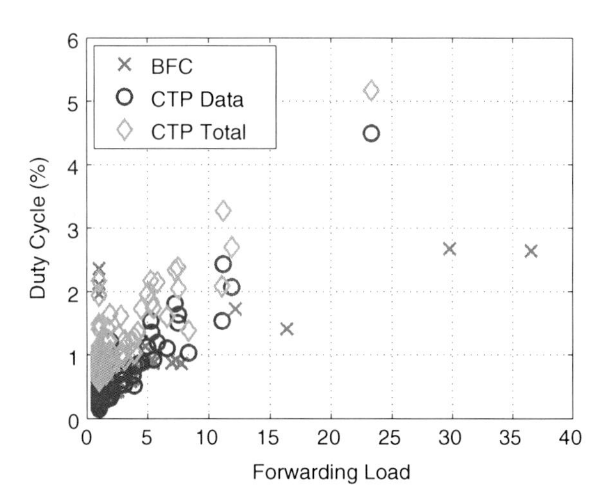

Fig. 12 Duty cycle versus forwarding load for the relays and the leaves

for forwarding only their own load (normally). Therefore, Fig. 12 depicts both node classes jointly. We observe that, in general, BFC penalizes some individual nodes. As hypothesized in [6], this may be due to a "rich gets richer effect". Nodes that are picked first as parents are eavesdropped first, which in turn increases their number of children and their forwarding load. This load balancing problem is not an intrinsic problem of LQE-free protocols. We believe that the opportunistic nature of ORW will naturally balance the load in the network because it can choose vastly different paths at each time (due to the presence of intermittent links). Analyzing the load balancing properties of ORW (among its many other LQE-free characteristics) is part of our future work.

4 Conclusions

Our evaluation indicates that LQE might have been overrated. Not using LQE results in longer routes with extra hops, but it also results in more conservative routing decisions, i.e. choosing individual links with a lower cost. The path length inefficiency caused by the lack of LQE is therefore amortized at the root; on top of that, no LQE means no control overhead, which translates into a reduction of the energy footprint of collection protocols between 40 and 60 % (for LPL wake-up interval of 1 s and inter-packet interval of 5 min). Lack of fairness posed by load imbalance is a clear drawback of LQE-free operation, though it can be avoided by abstracting away from the network topology and getting every node to receive every packet as in [5]. Our results, however, clearly show that load imbalance does not have sizable consequences in terms of energy consumption.

Acknowledgments We wish to thank our shepherd, Jason Gu, the anonymous reviewers, and the maintainers of publicly available testbeds. We gratefully acknowledge the TinyOS community and in particular all those researchers that have made their implementations available.

References

1. Gnawali, O., Fonseca, R., Jamieson, K., Moss, D., Levis, P.: Collection tree protocol. In: Proceedings of the 7th ACM Conference on Embedded Networked Sensor Systems, pp. 1–14. ACM, New York (2009)
2. Puccinelli, D., Haenggi, M.: Reliable data delivery in large-scale low-power sensor networks. ACM Trans. Sens. Netw. (TOSN) **6**(4), 28:1–28:41 (2010)
3. Moeller, S., Sridharan, A., Krishnamachari, B., Gnawali, O.: Routing without routes: The backpressure collection protocol. In: Proceedings of the 9th ACM/IEEE International Conference on Information Processing in Sensor Networks, pp. 279–290. ACM, New York (2010)
4. Landsiedel, O., Ghadimi, E., Duquennoy, S., Johansson, M.: Low power, low delay: opportunistic routing meets duty cycling. In: Proceedings of the 11th ACM/IEEE International Conference on Information Processing in Sensor Networks, pp. 185–196. ACM, New York (2012)
5. Ferrari, F., Zimmerling, M., Mottola, L., Thiele, L.: Low-power wireless bus. In: Proceedings of the 10th ACM Conference on Embedded Networked Sensor Systems, pp. 1–14. ACM, New York (2012)
6. Puccinelli, D., Zúñiga, M., Giordano, S., Marrón, P.J.: Broadcast-free collection protocol. In: Proceedings of the 10th ACM Conference on Embedded Networked Sensor Systems, pp. 29–42. ACM, New York (2012)
7. Doddavenkatappa, M., Chan, M.C., Ananda, A.L.: Indriya: a low-cost, 3d wireless sensor network testbed. In: Testbeds and Research Infrastructure. Development of Networks and Communities. Springer pp. 302–316 (2012)
8. Moss, D., Levis, P.: BoX-MACs: Exploiting Physical and Link Layer Boundaries in Low-Power Networking. Technical Report SING-08-00, Stanford University (2008)
9. Puccinelli, D., Gnawali, O., Yoon, S., Santini, S., Colesanti, U., Giordano, S., Guibas, L.: The impact of network topology on collection performance. In: Wireless Sensor Networks, pp. 17–32. Springer, Heidelberg (2011)
10. Dunkels, A., Osterlind, F., Tsiftes, N., He, Z.: Software-based on-line energy estimation for sensor nodes. In: Proceedings of the 4th Workshop on Embedded Networked Sensors, pp. 28–32. ACM, New York (2007)

11. Baccour, N., Koubaa, A., Mottola, L., Zúñiga, M.A., Youssef, H., Boano, C.A., Alves, M.: Radio link quality estimation in wireless sensor networks: a survey. ACM Trans. Sens. Netw. (TOSN) **8**(4), 34:1–34:33 (2012)
12. Alizai, M.H., Landsiedel, O., Link, J.Á.B., Götz, S., Wehrle, K.: Bursty traffic over bursty links. In: Proceedings of the 7th ACM Conference on Embedded Networked Sensor Systems, pp. 71–84. ACM, New York (2009)
13. Boano, C.A., Voigt, T., Noda, C., Römer, K., Zúñiga, M.: JamLab: Augmenting sensornet testbeds with realistic and controlled interference generation. In: Proceedings of the 10th IEEE International Conference on Information Processing in Sensor Networks, pp. 175–186 (2011)

Redundant Border Routers for Mission-Critical 6LoWPAN Networks

Laurent Deru, Sébastien Dawans, Mathieu Ocaña, Bruno Quoitin and Olivier Bonaventure

Abstract Sensor networks are gradually moving towards full-IPv6 architectures and play an important role in the upcoming Internet of Things. Some mission-critical applications of sensor networks will require a level of reliability that excludes the presence of single points of failure, as it is often the case today for the gateways connecting sensor networks to the Internet. In this chapter, we introduce RPL-6LBR, a 6LoWPAN border router that addresses mission-critical deployments through redundancy. The chapter discusses how existing standards may be leveraged to enable redundant border router synchronization, while identifying certain of their shortcomings. We also propose innovative network architectures incorporating multiple border routers, which deal with redundancy and node mobility without requiring any synchronization among the border routers. We implement the proposed RPL-6LBR in the Contiki operating system and report on this implementation through trials on a small-scale testbed and simulator. Our results open new possibilities for real-world wireless sensor networks requiring reliable border routers, and guide further standardization efforts in emerging technologies in support of the Internet of Things.

L. Deru · S. Dawans(✉) · M. Ocaña
CETIC, Rue des Frères Wright 29/3, B-6041 Charleroi, Belgium
e-mail: sebastien.dawans@cetic.be

L. Deru
e-mail: laurent.deru@cetic.be

M. Ocaña
e-mail: mathieu.ocana@cetic.be

B. Quoitin
Université de Mons, 20, place du Parc, B-7000 Mons, Belgium
e-mail: bruno.quoitin@umons.ac.be

O. Bonaventure
Université Catholique de Louvain, B-1348 Louvain-la-Neuve, Belgium
e-mail: olivier.bonaventure@uclouvain.be

K. Langendoen et al. (eds.), *Real-World Wireless Sensor Networks*,
Lecture Notes in Electrical Engineering 281, DOI: 10.1007/978-3-319-03071-5_20,

1 Introduction

In the foreseeable future, billions of low power wireless devices will be connected to the Internet. To make this Internet of Things (IoT) vision a reality, constrained devices are adopting standards-based solutions [1]. The IoT embraces the IPv6 networking architecture to fulfill these requirements. IPv6-based Wireless Sensor Networks (WSN) enable a diverse range of application domains from smart cities to building and home automation. As the IoT-enabling technologies mature, so do the requirements on interconnectivity, reliability and fault-tolerance.

A key element of any deployed IoT network is the border router (BR) that connects it to the Internet. The BR is a gateway between two different link-layer technologies, typically 802.15.4 on the WSN side, and Ethernet or Wi-Fi on the other. The BR also separates the control planes of two different routing domains. In the majority of current deployments, the BR is unique.

When deploying large IoT networks, several issues appear such as the high solicitation of nodes close to the BR, the single point of failure (SPoF) that the BR is, and interconnecting seamlessly with existing IPv6 networks. We argue that for reliable mission-critical networks, redundant BRs need to be deployed. With multiple BRs, the packets can be load-balanced across these BRs and their neighboring nodes if the BRs are deployed at different physical locations, spreading the energy consumption across the network and reducing the likelihood of bottlenecks during traffic bursts. Using multiple BRs also reduces the risk of unreachability of nodes if network partitioning occurs due to failures within the WSN. However, using multiple BRs raises new problems such as the need to synchronize them, how to recover from failures, how to handle node mobility and how to integrate them seamlessly into an existing IPv6 network.

In this chapter, we analyze how large IoT networks attached via redundant BRs can be deployed. Our work focuses on 6LoWPAN networks running RPL, the Routing Protocol for Low-Power and Lossy Networks (LLNs), as it is the primary routing protocol for IPv6 LLNs [2]. We analyze in detail how RPL networks support multiple BRs both in theory and in practice. We design a RPL 6LoWPAN Border Router (RPL-6LBR) solution supporting multiple BRs with node mobility and fault-tolerance without compromising the energy-efficient control mechanisms provided by RPL. Finally, we implement RPL-6LBR in the Contiki operating system and evaluate its performance.

2 Background

6LoWPAN and IPv6 operate on link layers that are fundamentally different: IPv6 link layers such as Ethernet are highly-available, provide high throughputs with low latencies and are more reliable than LLNs. As a consequence, control-plane protocols for IPv6 and 6LoWPAN networks do not share the same constraints. In this section, we describe how RPL allows for low-overhead networking in support of LLNs, and we further discuss how existing standards propose to interconnect IPv6 networks and 6LoWPAN.

2.1 RPL-based 6LoWPAN Networks

The Routing Protocol for LLNs (RPL) is a distance-vector routing protocol addressing the problem of scalable any-to-any routing in low-power IPv6 networks [2]. RPL routes upwards and downwards along a tree-like topology called Destination-Oriented Directed Acyclic Graph (DODAG), maintained via an adaptive beaconing mechanism. RPL provides good scalability with the network size because all the control traffic is focused in optimizing a single topology towards a single destination, rather than one route between all pairs of nodes in the network. It also favors data-driven link estimations, meaning it will not proactively waste energy attempting to optimize its DODAG unless data packets are routed, which is attractive in terms of power savings. The main innovation of RPL is that it brings bidirectional traffic patterns over lossy, multihop networks.

2.2 Interconnecting LLNs with IPv6

Figure 1 represents a typical network architecture integrating 6LoWPAN and IPv6 networks. The BR may act as a router and separate the IPv6 and 6LoWPAN networks in different subnets, or act as a bridge, handling both IPv6 and 6LoWPAN networks as a common subnet.

These architectures support end-to-end IP communication between any host and a sensor on the LLN. To achieve this goal, 6LoWPAN transmits IPv6 packets over IEEE 802.15.4 frames by adding an adaptation layer for header compression and fragmentation [3]. For IPv6 control plane mechanisms, like the Neighbor

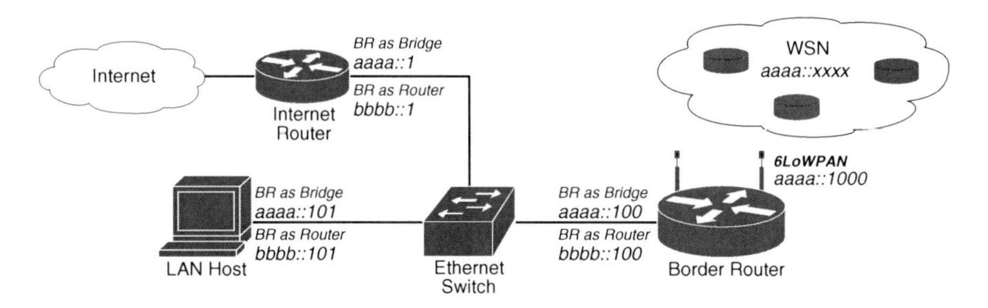

Fig. 1 Border Router in a typical IPv6-6LoWPAN environment

Discovery Protocol (NDP) [4], to operate on energy constrained and asymmetric networks, the IETF proposes the Neighbor Discovery Optimization for LLNs (6LoWPAN-ND) [5].

6LoWPAN-ND aims to support both proposed approaches to routing in 6LoWPAN networks: route-over and mesh-under modes. However, 6LoWPAN-ND does not fully encompass the problem of interconnecting 6LoWPAN-based WSNs to the Internet, only interconnection by routing is explored.

To allow transparent communication as if the LLN is part of the IPv6 subnet, the Neighbor Discovery Proxy Gateway for 6LoWPAN-based WSNs (6LP-GW) provides bridging capabilities and allows an external host to discover and connect to sensors using NDP and protocol conversion between NDP and 6LoWPAN-ND [6]. However, that proposal does not cover the 6LoWPAN route-over scenario and its interaction with RPL.

2.3 Dealing with Mobility and Redundancy

Dynamic routing protocols like RPL are designed to adapt their routing strategies against varying channel conditions. As a side effect, an immobile sensor can appear to be mobile. RPL has built-in mechanisms to support inter-LoWPAN mobility thanks to the multi-instance support and the capability to switch from one DODAG to another in the same instance.

6LoWPAN-ND does not have any requirement on inter-LoWPAN mobility in route-over networks or on multi-BR support. This is out of scope of [5]. However, [5] requires that either sensors register their address towards all of the known 6LBRs or that an out-of-scope mechanism synchronize those 6LBRs. Again, no current 6LoWPAN-ND implementation proposes any synchronization mechanisms. Thus, inter-LoWPAN mobility and redundancy is severely constrained as a WSN node moving from one DODAG to another will see the subnet prefix changing and therefore will get a new global address. If each LLN is configured with its own prefix, Mobile IPv6 or one of its variants could be implemented [7].

3 Design of a RPL-Compatible 6LBR

In order to tackle the shortcomings identified in the previous sections, we propose to define the routing and mobility requirements for multi-BR RPL-based networks as follows:

1. The BR must be compatible with RPL in a route-over mode.
2. Multiple BRs must be able to co-exist in the same infrastructure, handling contiguous or disjoint LLNs.
3. Only one BR at a time should be responsible for a given 6LoWPAN host.

4. LLNs connected to different BRs should be able to share the same prefix.
5. A 6LoWPAN host should be able to maintain its global IPv6 address when switching between BRs that share a common prefix.
6. Multiple WSNs and an adjacent Ethernet segment can share the same prefix.

These requirements form the basis of our RPL-6LBR design, described below.

3.1 ND Proxy for RPL-6LBR

If the LLNs share the same prefix, one could naïvely imagine that it is enough to forward packets between the two interfaces. However, interoperability problems arise as the IPv6 hosts will use NDP to resolve addresses whereas the hosts on the WSN side will use routes and default routes provided by RPL to communicate. The RPL-6LBR must provide an adaptation mechanism, handling NDP requests and converting NDP configuration options to RPL options. The concept of ND-Proxy is not a new one, a similar concept has been defined in RFC 4389 ND Proxy [8], but it is mainly targeted at proxifying wireless or PPP based bridges and not applicable as such for RPL-6LBR. ND Proxy has also been implemented for adapting 6LoWPAN-ND with NDP [6], and proposed for RPL [9]. In our case we propose to bridge a RPL network with a NDP network; the information gathered through RPL messages is used for the ND proxy functionalities and the configuration parameters received through NDP configure the RPL network.

3.2 Mobility Support for Redundant 6LBRs

Aggregating several WSNs and allowing sensor mobility across them is not currently defined in 6LoWPAN-ND or RPL. Following the description of the possible LLNs topologies, aggregation of LLNs can be done at different levels, depicted in Fig. 2. It should be noted that redundancy can be seen as an extreme case of WSN aggregation, where the two WSNs are identical.

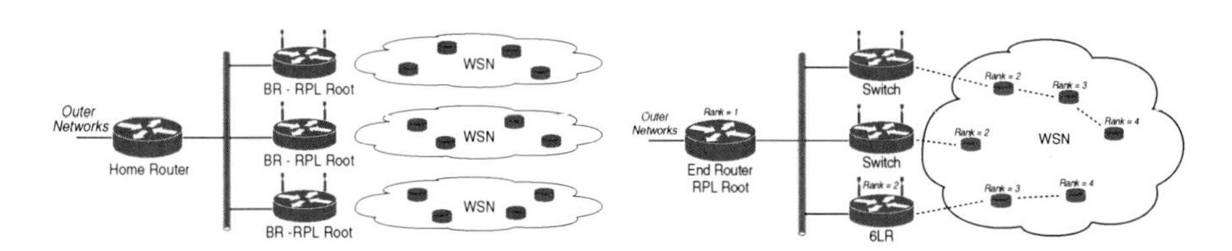

Fig. 2 *Left* Routing level aggregation, each BR is a RPL Root with its own DODAG. *Right* RPL level aggregation, the RPL Root runs on the backbone

Aggregating WSNs by synchronizing RPL-6LBR Aggregating WSNs using ND Proxy based RPL-6LBR sharing the same prefix is feasible but requires the synchronization of the different RPL-6LBRs to allow sensor mobility. Otherwise, several RPL-6LBRs would perform ND Proxy for the same WSN node, which is undesirable. In this chapter we rely on the Neighbor Advertisements to synchronize the ND Proxies of the connected RPL-6LBR.

Aggregating WSN using an external DODAG root RPL integrates sensor mobility deeply, so an innovative solution we propose is to create a unique DODAG across multiple BRs encompassing all the WSN nodes. This unique DODAG is managed by an External DODAG Root connected to the BRs through an Ethernet backbone.

4 Implementation

We provide an implementation of our RPL-6LBR in the latest development version of the Contiki operating system.[1] We support multiple types of RPL-6LBR platforms: the RapsberryPi with a custom 802.15.4 radio, the BeagleBone, the Econotag running Contiki natively and a standard Linux PC.

To enable multiple BR support in Contiki, we implement an ND Proxy as designed in Sect. 3.1 to adapt IPv6/NDP and 6LoWPAN/RPL. We also implement prefix distribution by leveraging the Prefix Information Option (PIO) of the RPL DIO broadcast messages combined with a global reset of the DIO Trickle timer. Due to space limitations we refer the reader to the online documentation for implementation details.

5 Evaluation

We evaluate RPL-6LBR in different scenarios with a Python-based unit test framework, provided as part of the RPL-6LBR open-source repository.

As our implementation targets real deployments, the results in this chapter are partially made up of tests on the RaspberryPi, interfaced with real sensor nodes of our testbed: 20 nodes spread over a 500 m^2 office space (10 TelosB, 10 Zolertia Z1). In order to explore certain parameters in a controlled manner, we also parallelized a large number of test instances on Linux virtual machines, interfacing the same RPL-6LBR with a virtual WSN using the COOJA simulator with TelosB emulated sensor devices.

All sensor modules run a Contiki application over the ContikiRPL implementation using the most commonly used configuration: the Minimum Rank Hop Objective Function (MRHOF) with ETX [10] as the routing metric and with RPL's storing mode for downwards routing. We control the network size, topology, RPL DAG stabilization times and types and amount of cross-traffic. All runs consist in analyzing

[1] http://cetic.github.com/6lbr. tag 6blr-1.1.0. 2013 (licence: BSD).

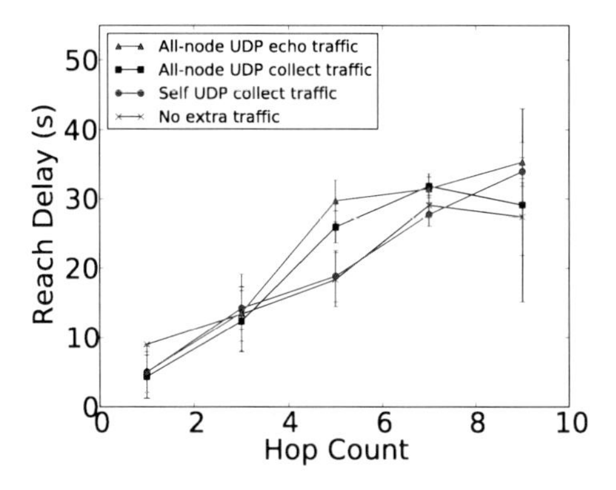

Fig. 3 Time of propagation of new prefix versus hop count

the effects of cross-traffic, node mobility and prefix reconfiguration on bidirectional application traffic between a host on the Ethernet network and a constituant of the WSN network, as far down the RPL DAG as possible. Each testcase was executed 10 times, with different random seeds for the simulated ones.

Prefix distribution We measure the lapse of time during which a particular host on the WSN is unreachable on a new prefix while this prefix is propagated on a running network. Figure 3 shows the average duration of unreachability versus the distance from the RPL Root for different traffic patterns. The main result is that there is little effect of the cross-traffic, prefix distribution being only dependant of the beaconing mechanisms of RPL.

Inter-WSN routing We next evaluate inter-WSN routing between two RPL-6LBRs set up in both SmartBridge and Switch modes, handling physically disjoint WSNs. Ping messages are exchanged between elements of disjoint DODAGs (garanteed in simulation), each roughly three hops away from its root. In SmartBridge mode, the Round Trip Times (RTT) of pings average $0.62\,\text{s} \pm 0.12\,\text{s}$, which is slightly shorter than the ones in the Switch mode measured as $0.67\,\text{s} \pm 0.12\,\text{s}$ due to the extra hop that is induced in Switch mode.

Node mobility On Fig. 4, we observe the behavior of RPL-6LBR in the presence of a mobile node switching between DODAGs. We see that the performance is dependant of RPL's link metric estimation. High data collection rates update the mobile node's ETX rapidly enough to allow it to select a member of the secondary DODAG as preferred parent in all cases. On the other hand, low data collection rates do not all result in the reachability of the node within a 600 s limit we imposed, and the trials that do converge exhibit a longer convergence time on average.

Border router outages In the case of BR outages, the effects of data-traffic is amplified (Fig. 5). A BR failure orphans the entire sub-DODAG originally handled by the BR and requires nodes closest to the alternate DODAG to update their link ETX first, then join the DODAG and announce their ranks to neighboring orphans via DIOs. The remaining orphans of the previous DODAG gradually attach themselves to the new DODAG in a hop by hop manner.

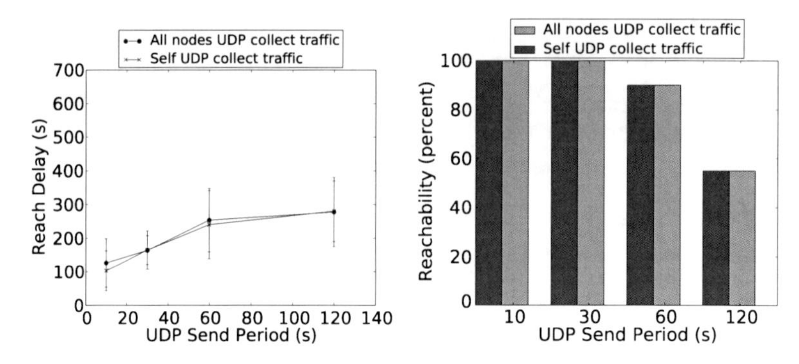

Fig. 4 Mobile node trials. The success rate and unreachability duration is independent of whether or not the other nodes of the network update their link metrics with periodic unicasts

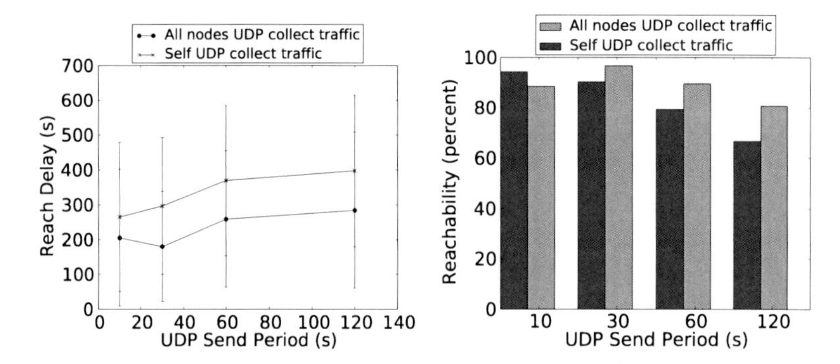

Fig. 5 BR outage trials. The success rate and unreachability delay are improved when all elements of the network accelerate their ETX updates as they emit periodic UDP messages

A close look at our test results reveals that routing loops appeared among the orphaned elements of the failed cases, which prevents ETX from rising since unicasts keep getting acknowledged within the routing loop. This is still an open problem in ContikiRPL, as it is capable of detecting routing loops using the IPv6 Hop-by-Hop Option for RPL [11] but does not implement any local repair.

6 Discussion

Our evaluation covers functional and performance aspects of the RPL-6LBR implementation. We started by showcasing the prefix distribution functionalities, which is an important deployment aspect for real-world WSNs. Our results shows that the principle of a RPL ND-Proxy to interconnect IPv6 networks and RPL networks is sound and reliable. The synchronization mechanism, based only on existing standards and messages, is proved to be functional and allows fast transition of one sensor node from one WSN to another.

Performance-wise our measurements show a heavy influence of ETX-based link estimation on RPL hosts' reactiveness to change, in both mobility and BR outages scenarios. Low-overhead link estimation is desirable for long-running battery-operated sensor networks, and it is important to note that the solution we propose does not compromise this design.

7 Conclusion

This work introduces a new 6LoWPAN border router design and implementation which leverages existing standardization efforts while complying with the reactive link estimation mechanisms of RPL. We also propose solutions for redundant BR deployments and assess RPL-6LBR in the presence of mobile nodes and BR failures. The results show that RPL-6LBR is viable for redundant BR deployments, but suggest that a different choice of Objective Function would be needed for critical systems.

In future work, multiple RPL instances will accommodate different types of routing QoS to harness the flexibility of RPL. We will also address larger WSN deployments by extending our testbed and integrating multiple RPL-6LBRs on larger, existing testbeds and real-world deployments.

Acknowledgments This work was partly funded by the Walloon Region under the First-DoCA funding number 1017211. The authors would also like to thank Maxime Denis from UMons for his valuable contributions.

References

1. Ishaq, I.: IETF standardization in the field of the internet of things (IoT): a survey. J. Sen. Actuator Netw. **2**(2), 235–287 (2013) Issn: 2224–2708. doi:10.3390/jsan2020235
2. Winter, T., Thubert, P., RPL Author Team: IPv6 Routing Protocol for Low power and Lossy Networks. March 2012
3. Montenegro, G., et al.: Transmission of IPv6 Packets over IEEE 802.15.4 Networks. Internet proposed standard RFC 4944, September 2007
4. Narten, T., et al.: Neighbor Discovery for IP version 6 (IPv6). RFC 4861. IETF, September 2007
5. Shelby, Z., et al.: Neighbor discovery optimization for IPv6 over low-power wireless personal area networks (6LoWPANs). Standard Track 6775, IETF, November 2012
6. Maqueda Ara, L.: Neighbor Discovery Proxy-Gateway for 6LoWPANbased Wireless Sensor Networks. MA thesis, KTH Information and Communication Technology (2011)
7. Obradovic, A., Nikolic, G.: Overview of mobility protocols features for 6LoWPAN. In: YUINFO, Wiley 2012
8. Thaler, D., Talwar, M., Patel, C.: Neighbor Discovery Proxies (ND Proxy). Experimental RFC 4389, April 2006
9. Thubert, P.: 6LoWPAN Backbone Router. IETF Draft, 2013 http://datatracker.ietf.org/doc/draft-thubert-6lowpan-backbone-router/
10. De Couto, D.S.J., et al.: High-throughput path metric for multi-hop wireless routing. In: Proceedings of the International Conference on Mo- bile Computing and Networking (ACM MobiCom) pp. 134–146. San Diego, CA, USA, (2003)
11. Hui, J., Vasseur, J.P., The RPL Option for Carrying RPL Information in Data-Plane Datagrams. RFC 6553, March 2012

Using Directional Transmissions and Receptions to Reduce Contention in Wireless Sensor Networks

Ambuj Varshney, Thiemo Voigt and Luca Mottola

Abstract Electronically Switched Directional (ESD) antennas allow software-based control of the direction of maximum antenna gain. ESD antennas are feasible for wireless sensor network. Existing studies with these antennas focus only on controllable directional *transmissions*. These studies demonstrate reduced contention and increased range of communication with no energy penalty. Unlike existing literature, in this chapter we experimentally explore controllable antenna directionality at both sender and receiver. One key outcome of our experiments is that directional transmissions and receptions together considerably reduce channel contention. As a result, we can significantly reduce intra-path interference.

1 Introduction

Electronically switched directional (ESD) antennas allow software-based control of the direction of the maximum antenna gain. ESD antennas bring spatial diversity to wireless applications, and have been shown feasible for real world sensor networks. Previous work has studied the impact of introducing controllable directionality at the sender nodes only. These studies demonstrate improvements in network performance because of reduced contention [1] and increased range of communication [2, 3]. There is, however, no experimental evidence about performance

A. Varshney (✉) · T. Voigt
Uppsala University, Uppsala, Sweden
e-mail: ambuj.varshney@it.uu.se

T. Voigt · L. Mottola
SICS Swedish ICT, Kista, Sweden
e-mail: thiemo@sics.se

L. Mottola
Politecnico di Milano, Milan, Italy
e-mail: luca@sics.se

K. Langendoen et al. (eds.), *Real-World Wireless Sensor Networks*,
Lecture Notes in Electrical Engineering 281, DOI: 10.1007/978-3-319-03071-5_21,

improvements brought by introducing controllable antenna directionality at both sending and receiving nodes.

Directional transmissions alleviate contention by conveying radiated power in the intended direction of communication. Nevertheless, antennas are reciprocal in nature, i.e., they have similar receiving and sending patterns [4]. This suggests directional receptions enabled by these antennas could, for example, help alleviate channel contention from nearby nodes. Increased contention for the channel leads to higher packet loss, increased latency, and decreased throughput resulting in decreased lifetime of sensor network applications. Introducing directional reception could further alleviate contention by attenuating the signal at the receivers from nodes in unintended directions of communication.

We build a number of SPIDA ESD antennas [2] for our experiments in this chapter. We evaluate these antennas as receivers and observe similarity in sending and receiving patterns. We experiment with these antennas arranged in a rectangular grid and a linear chain of nodes. Our experiments confirm that directional transmissions and receptions reduce channel contention. Our experiments also suggest that we can significantly reduce intra-path interference in linear networks, a problem experienced in high-throughput protocols such as Flush [5] and PIP [6]. Finally, we demonstrate that by exploiting directional transmissions and receptions and the capture effect, simultaneous communication flows between multiple sender-receiver on one wireless channel only are possible. In contrast to other protocols such as Strawman [7] that reduces the contention by distributing transmissions in time, our approach tackles the problem in space.

The key contribution of this chapter is to confirm that directional transmission and reception together indeed significantly reduce channel contention and intra-path interference. The rest of the chapter unfolds as follows: Sect. 2 provides a brief background on ESD antennas and verifies that the prototypes we build exhibit a directional behavior. In Sect. 3 we report on our experiments demonstrating how exploiting directional transmissions and receptions can reduce contention and intra-path interference. Section 4 places our results in perspective against existing literature and concludes the chapter.

2 Electronically Steerable Directional Antennas

The SICS Parasitic Interference Directional Antenna (SPIDA) is based on the concept of *Electrically-Switched Parasitic Element*. Nilsson designed SPIDA for low powered wireless-sensor networks [2]. SPIDA has six parasitic elements surrounding a quarter wavelength monopole antenna. The parasitic elements can be individually grounded or isolated. When all parasitic elements are isolated, the antenna is configured in omni-directional mode. When all elements are grounded except one, the direction of maximum antenna gain points towards the direction of the isolated element. Encouraged by results obtained with the SPIDA antenna [1, 2, 8], we construct and use SPIDA antennas for our experiments.

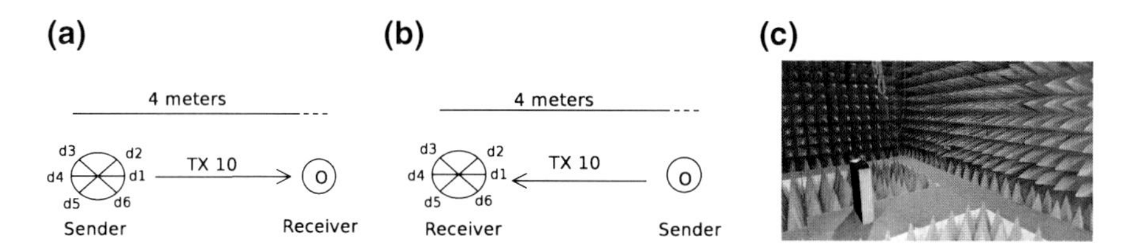

Fig. 1 **a** Sending pattern experiment. **b** Receiving pattern experiment. **c** Experiment setup in anechoic chamber. Experimental setup used to demonstrate similar sending and receiving pattern of SPIDA antennas. d1–d6 indicate possible SPIDA directions. d7 indicates omni-directional configuration. O indicates a probe node with omni-directional antenna

We evaluate the antenna prototypes we build in terms of the ability to control the direction of maximum antenna gain. We further evaluate the receiving behaviour and sending behaviour of SPIDA antennas to show reciprocity in sending and receiving patterns of SPIDA antennas.

Our experimental setup consists of Tmote sky nodes equipped with the antennas as shown in Fig. 1. We perform the experiment in an anechoic chamber to reduce the effect of multi-path and external interference. As a sender, the SPIDA-equipped node is configured to broadcast packets containing the sending direction, sequence number, and transmit power at an inter-packet interval (IPI) of $\frac{1}{2}$ second. Even though we are in an anechoic chamber, we chose this IPI since it usually prevents successive packet loss due to link burstiness [9] and our experiments in the next section are not performed in the chamber. We reconfigure the direction of the maximum antenna gain in a round robin manner sending ten packets in one direction before switching direction. When receiving a packet the probe node logs RSSI, antenna configuration, sequence number and node id onto onboard flash. In the second experiment we observe the receiving pattern. The roles of the SPIDA-equipped node and probes are reversed, keeping all other parameters the same. The node equipped with the omnidirectional antenna broadcasts beacon messages. The receiver with the SPIDA antenna stores RSSI, receiving direction, and sequence number onto onboard flash.

Figure 2 shows the result of our experiments. The figure depicts the mean RSSI of the received packets for five different SPIDA antennas used as sender and as receiver in the first and second type of experiment, respectively. The error bars show the standard deviation across the antenna prototypes. The graph shows that we can control the direction of the maximum antenna gain with the received signal strength being the highest when the antenna is configured in the direction of node 1. Configuring the direction of the maximum antenna gain away from the node leads to a decrease in signal strength of the received packets, with direction 3 and 5 being the worst performing directions. This is consistent with earlier results [8]. The more interesting result is the large difference in signal strength between the best direction (direction 1) and the worst direction (direction 3). We also observe as expected, a close resemblance in the sending and receiving patterns of the SPIDA antenna, which demonstrates the antenna's reciprocity.

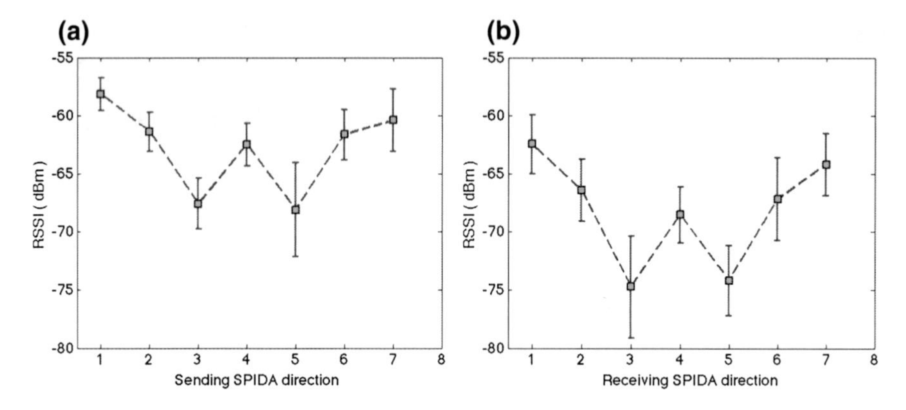

Fig. 2 **a** Sending pattern. **b** Receiving pattern. Mean RSSI using SPIDA antenna as sender and as a receiver. Changing direction of maximum gain has significant effect on RSSI. *The receiving pattern looks very similar to the sending radiation pattern suggesting reciprocity*

3 Experimental Evaluation

In this section, we show that directional transmissions and receptions reduce channel contention. We also show that this allows nodes in a linear network to communicate simultaneously on the same wireless channel.

3.1 Basic Experimental Setup

We arrange the nodes according to Figs. 3 and 4, for two topologies that we call rectangular and linear. These topologies allow us to exploit directionality at the sender and the receiver. The nodes are arranged with direction 1 of the sender pointing towards direction 1 of the receiver antenna in line of sight. Henceforth, configuring SPIDA antenna to directional mode means that we configure the direction of the maximum antenna gain towards direction 1.

A sensor node with an omnidirectional antenna broadcasts beacon messages at transmit power TX 31 (approximately 0 dBm). We use a higher transmit power to ensure that beacon messages are received by all intended receivers independent of their antenna configuration. When receiving a beacon message, the sender and receiver nodes configure the direction of the maximum antenna gain to directional or omnidirectional mode. As the experiments are performed indoors, with nodes separated by a few meters, we use the lower transmit power TX7–TX9 for sender nodes. We use the same transmit power for all nodes in the rectangular topology. In the linear topology nodes have incrementally higher transmit power settings according to their placement in the chain. This is required due to the short distances between the nodes in the chain and not needed when the distances between them are larger.

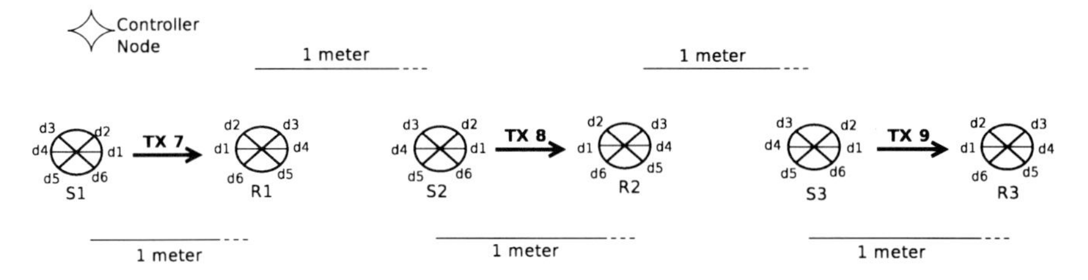

Fig. 3 Experimental setup for linear communication. *Arrow* directions denote the paired nodes. TX7, TX8, TX9 indicate the senders' output power

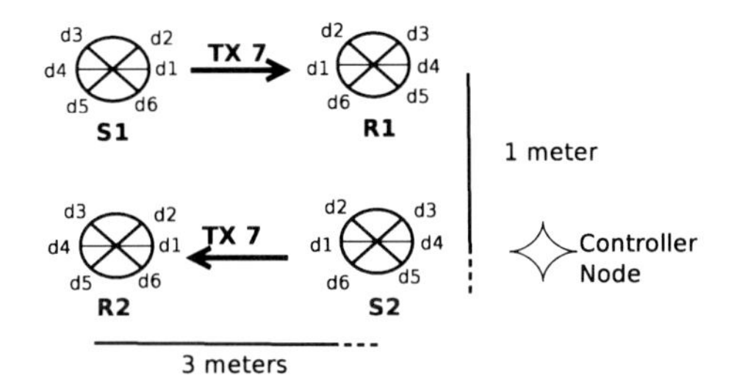

Fig. 4 Experimental setup for investigating channel contention. Nodes are arranged in a rectangular topology. *S* indicates sender and *R* receiver node

Again, we set the inter packet interval to $\frac{1}{2}$ second to prevent successive packet losses due to link burstiness. To prevent interference from IEEE 802.11 networks, we use the IEEE 802.15.4 channel 26 in our experiments.

3.2 Alleviating Channel Contention

We investigate if directional transmissions and receptions can alleviate channel contention. We establish communication between paired nodes, i.e., S1–R1, S2–R2 and S3–R3 (linear topology only). Our goal is to show that we can alleviate contention from unpaired nodes.

The nodes are arranged as discussed in the previous section. The sender nodes broadcast packets with sender node id and antenna configuration after receiving the beacon message. In these experiments we introduce a delay before we trigger the senders' broadcasts to prevent collisions of packets from different sender nodes. The receiver node logs RSSI, sender node id, as well as sending and receiving antenna configuration onto the onboard flash. In the experiments we collect roughly 7,000 packets.

Figure 5 shows the results of the experiment with nodes arranged in the rectangular grid, Fig. 6 with nodes arranged linearly. In the graphs, we plot the mean received RSSI of packets for the different antenna configurations. When nodes are arranged in

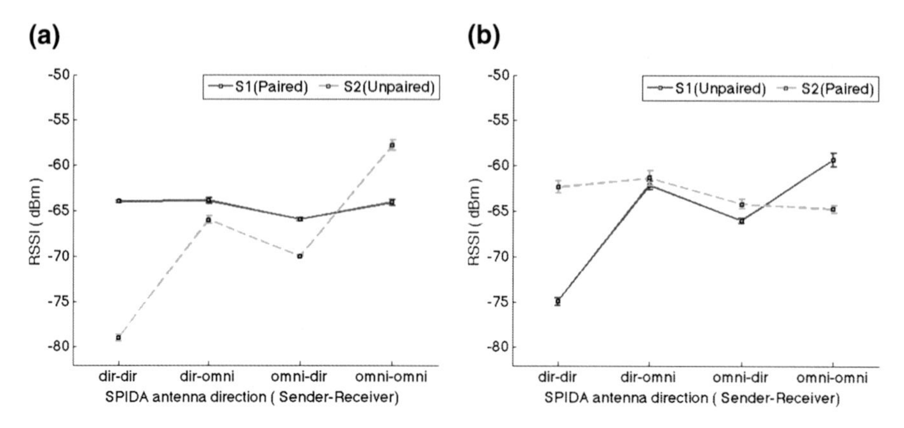

Fig. 5 **a** Node 1 (R1). **b** Node 2 (R2). Mean received RSSI of packets for the antenna configuration on the X-axis. Nodes are arranged in a rectangular grid. *Introducing directionality reduces the RSSI from the nearby unpaired node*

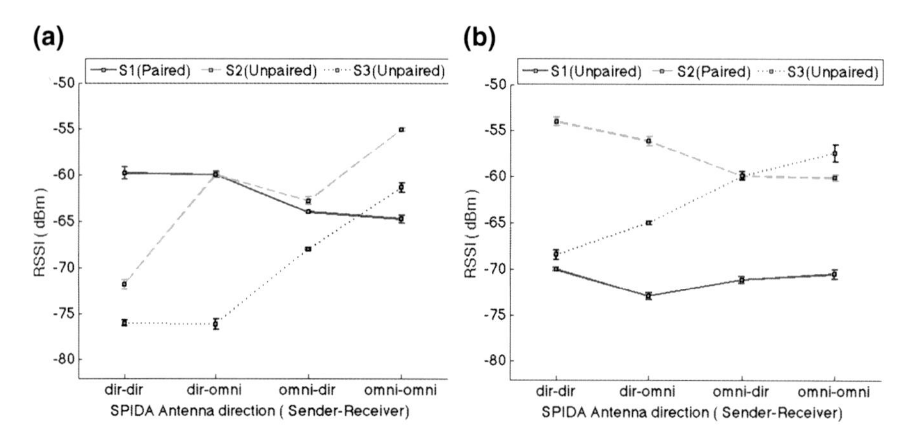

Fig. 6 **a** Node 1(R1). **b** Node 2(R2). Mean RSSI of received packets from different sender nodes with nodes arranged linearly. *Configuring the direction of the SPIDA Antenna helps to reduce intra-path interference*

the rectangular grid the RSSI of packets sent by the unpaired node is the highest. This is because of the proximity of the unpaired node and the omnidirectional configuration. However, as sender and receiver are configured to directional mode, the RSSI of packets sent by the unpaired node is reduced. The graphs shows that configuring only the sender or the receiver to directional mode has significantly less effect than configuring both to directional mode. We see a 21 dB (Fig. 5a) and a 15.6 dB (Fig. 5b) difference in RSSI for packets sent by the unpaired node between omnidirectional and directional configuration. In directional mode the RSSI of packets sent by the paired sender is the highest confirming that directionality at both sender and receiver is key to alleviate channel contention.

Similar to the experiment with the rectangular topology, we expect directional transmissions and receptions to alleviate contention when nodes are arranged in the linear topology as shown in Fig. 3. Further, we expect that configuring directionality should attenuate signals from S2 and S3 for receiver node R1, and from S3 for node R2. We do not expect directional transmissions and receptions to alleviate contention for receiver node R3, as the direction of communication of S3 is the same as of S1 and S2. Also, transmissions from S2 interfere with transmissions from S1 at R1. Similarly S3 interferes with transmissions from S2 at R2. This interference is similar to intra-path interference [10].

Figure 6a depicts the RSSI for packets received from different sender nodes at receiver node R1. The graph shows that in omnidirectional mode the RSSI for packets from S2 is the highest since S2 transmits at higher transmit power. As we put both sender and receiver nodes to directional configuration, we are able to attenuate the RSSI of the packets R1 receives from both S2 and S3 significantly, with the maximum effect when both sender and receiver are configured to directional mode. We observe 16.8 and 14.7 dB difference in RSSI of packets sent by S2 and S3 in omnidirectional and directional configuration. In the directional configuration, the RSSI of the paired sender is higher. The similar behaviour can be seen for node R2. The graphs shows an 11 dB difference in RSSI for packets sent by S3. This confirms our finding that directional transmissions and receptions together significantly alleviate channel contention, and suggests that intra-path interference could be significantly reduced with directional transmissions and receptions.

3.3 Simultaneous Communication Flows

In this section, we build upon the results obtained in previous section and show that directional transmissions and receptions make it possible to establish communication flows between nodes on the same wireless channel. We demonstrate this by forcing simultaneous communication between paired nodes on the same wireless channel in the following experiments exploiting the capture effect [11].

The experimental setup is similar to the one in the previous section. To allow packets from different sender nodes to collide, we remove the delay introduced after the reception of the beacon message that triggers the sender to broadcast a packet. This causes the sender nodes to broadcast the packet at the same time. We have seen in the earlier experiment that directional transmissions and receptions ensure the signal strength of packet sent from paired sender node are the highest. The results also show that the difference in RSSI between the packets is >3 dB, which is the co-channel interference tolerance level of the CC2420. In this experiment, we expect because of the capture effect to receive only the packet with the higher RSSI from the paired sender node even in presence of other concurrent packet transmissions from the unpaired sender nodes.

We collect around 30,000 packets for both topologies. Figure 7a, b show the results when nodes are arranged in a rectangular grid, and Fig. 7c, d show the results when nodes are arranged linearly. In the graph, the bar plots of some sender nodes are

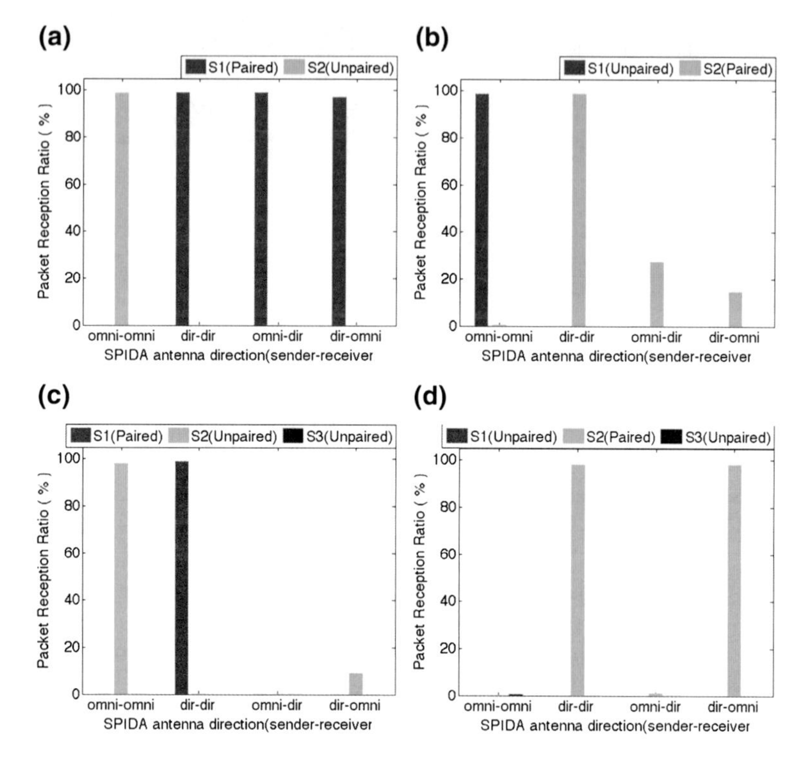

Fig. 7 **a** Node 1 (R1). **b** Node 2 (R2). **c** Node 1 (R1). **d** Node 2 (R2). Packet reception ratio (PRR) for receiver node 1 and 2. Nodes arranged in rectangular topology for **a** and **b**. Nodes arranged in linear topology for **c** and **d**. *A high PRR is observed from paired nodes when both sender and receiver are configured to directional mode*

not visible because the PRR is zero or close to zero. The graph clearly shows that configuring both the sender and receiver to directional mode allows us to establish communication between paired nodes with high packet reception ratio (PRR). Figure 7b, c and d show that configuring only the sender or the receiver to directional mode results in lower PRR. This is expected since the difference in RSSI between packets is close to 3 dB or less in the figures in the previous section. These experiments suggest that we can establish simultaneous communication flows on the same wireless channel using directional transmissions and receptions.

4 Discussion and Conclusion

Our experiments suggest that directional transmissions and receptions alleviate contention and can reduce intra-path interference in a linear network. Woo and Culler have shown that intra-path interference is a problem for reliable delivery of data in multi-hop wireless networks that is hard to avoid [10]. This is aggravated for high goodput bulk data transmission protocols such as Flush [5]. Therefore, protocols like PIP use channel diversity to avoid intra-path interference and improve end-to-end throughput [6]. Since there are only two IEEE 802.15.4 channels that do not overlap with the frequencies used by WiFi, channel diversity may require the use of channels that are interfered by WiFi.

The results of our experiments suggest that using directional transmissions and receptions we can avoid intra-path interference without using multiple channels. The latter also helps decrease protocol complexity and opens the possibility of high goodput multi-hop paths using a single wireless channel.

Acknowledgments This work has been supported by the WISENET center at Uppsala University.

References

1. Mottola, L., et al.: Electronically-switched directional antennas for wireless sensor networks: a full-stack evaluation. In: IEEE SECON (2013)
2. Nilsson, M.: Directional antennas for wireless sensor networks. In: Scandinavian Workshop on Wireless Adhoc, Networks (2009)
3. Giorgetti, G., et al.: Exploiting low-cost directional antennas in 2.4 GHz IEEE 802.15.4 wireless sensor networks. In: European Microwave Week (EuMW'07) (2007)
4. Balanis, C.A.: Antenna Theory: Analysis and Design. Wiley (2012)
5. Kim, S., et al.: Flush: A reliable bulk transport protocol for multihop wireless networks. In: ACM SenSys (2007)
6. Raman, B., et al.: Pip: a connection-oriented, multi-hop, multi-channel tdma-based mac for high throughput bulk transfer. In: ACM SenSys (2010)
7. Österlind, F., et al.: Strawman: resolving collisions in bursty low-power wireless networks. In: ACM IPSN (2012)
8. Öström, E., et al.: Evaluation of an electronically switched directional antenna for real-world low-power wireless networks. In: REALWSN (2010)
9. Srinivasan, K., Kazandjieva, M., Agarwal, S., Levis, P.: The beta factor: measuring wireless link burstiness. In: ACM SenSys (2008)
10. Woo, A., Culler, D.: A transmission control scheme for media access in sensor networks. In: ACM MobiCom (2001)
11. Leentvaar, K., Flint, J.: The capture effect in FM receivers. IEEE Trans. Commun. **24**(5), 531–539 (1976)

Part V
Energy

Energy Parameter Estimation in Solar Powered Wireless Sensor Networks

Mustafa Mousa and Christian Claudel

Abstract The operation of solar powered wireless sensor networks is associated with numerous challenges. One of the main challenges is the high variability of solar power input and battery capacity, due to factors such as weather, humidity, dust and temperature. In this article, we propose a set of tools that can be implemented onboard high power wireless sensor networks to estimate the battery condition and capacity as well as solar power availability. These parameters are very important to optimize sensing and communications operations and maximize the reliability of the complete system. Experimental results show that the performance of typical Lithium Ion batteries severely degrades outdoors in a matter of weeks or months, and that the availability of solar energy in an urban solar powered wireless sensor network is highly variable, which underlines the need for such power and energy estimation algorithms.

1 Introduction

The convergence of communication, computation and sensing into ever smaller platforms paves the way to a completely new set of monitoring and automation applications in cities. Such applications usually rely heavily on *wireless sensor networks* (WSNs), which consist of, an ad-hoc of network of sensors transmitting its data to one or multiple gateways for processing. Given the decreasing costs of electronics, WSNs are expected to have more and more importance in tomorrow's cities, with

M. Mousa (✉) · C. Claudel (✉)
King Abdullah University of Science and Technology,
Thuwal, MK 23955-6900, Saudi Arabia
e-mail: mustafa.mousa@kaust.edu.sa

C. Claudel
e-mail: christian.claudel@kaust.edu.sa

K. Langendoen et al. (eds.), *Real-World Wireless Sensor Networks*,
Lecture Notes in Electrical Engineering 281, DOI: 10.1007/978-3-319-03071-5_22,

applications including building monitoring [1], and urban sensing (traffic, pollution, lighting).

A significant portion of the global cost of a wireless sensor network is the installation, specially if dedicated power and/or data lines need to be installed. To reduce costs, it is desirable for the sensor network to be self-powered. Battery power can be used for ultra-low power applications (such as smart parking or traffic flow monitoring), though the wireless range becomes severely limited, which imposes additional relays to be installed, increasing costs. Harvesting some energy from the environment is possible, and allows increased sensing, communication and computational capabilities. Among all possible ambient energy sources, solar power is usually the most practical and reliable source in lower latitudes. However, there are some significant challenges associated with operating a solar powered WSN, particularly when the average solar power obtained by the solar panels is not considerably larger than the average current draw, which is the case for instance in traffic sensing applications (or other high power sensing applications).

In this article, we present the results of a medium scale experiment involving 34 nodes deployed in an urban environment over a three month period. The nodes are powered both by a solar panel and by a rechargeable Lithium ion battery. This experiment shows that the power availability and energy storage parameters are highly variable, and that node-level energy management is needed. We identified two main areas of focus to tackle the energy estimation and forecast problems:

- Battery capacity and condition estimation. Our experimental results show that the condition of the batteries is highly dependent on discharging patterns and environmental effects. In particular the remaining battery capacity, which is an important decisional factor for optimizing the operation of a wireless sensor network, is highly variable.
- Solar power estimation. Our experimental results show that even if the nodes are deployed in the same geographical area, the available solar power is highly variable. This is of considerable importance for sensing applications, since the sensing duty cycle of nodes can be dynamically adjusted in relation to their available and forecasted energy.

The node power and energy parameters are an input to any energy aware sensing and communication scheme [2–5]. Therefore, we need to estimate these parameters for efficient operations. The focus of the present article is indeed to introduce a suite of energy parameters estimation methods that can all be implemented in typical WSN microcontrollers.

2 Related Work

Energy monitoring is an active research field in WSNs, for instance with the work of [6] in energy monitoring. One of the difficulties arising with energy estimation is the absence of dedicated monitoring systems in most commercial platforms, which can require the development of dedicated hardware and software [7, 8]. In [9], the Li-Ion capacity depletion/discharge is modeled to predict future energy availability, though some critical parameters (battery capacity) are not dynamically estimated. Some researchers [10] consider some electrochemical phenomena, such as rate capacity characteristics, charge recovery and thermal effects, which can play a role in governing the selection of sensing parameters. In this chapter, our objective is to analyze the energy patterns in nodes subjected to a external source, in a harsh environment (humidity, dust, temperature), and to propose computationally efficient methods for estimating energy and power parameters dynamically. The resulting algorithms are to be implemented in the nodes themselves, which have limited computational capabilities.

3 Experimental Work

3.1 System

Our proposed system consists of an heterogeneous wireless sensor network for monitoring traffic flow in cities using both fixed and mobile sensor data. The specificity of this sensor network lies in its operation: the traffic state estimates are directly computed by the wireless sensor network, and are forwarded to an output database. This distributed computing based operation allows the system to be cheaper and simpler to deploy, since no costly redundant estimation servers and input databases would be required. It would also enhance user privacy, since user data would remain local (only traffic estimates would be global).

The platform chosen for this prototype is the `Libelium Waspmote`, which is a commercial sensing platform derived from the `Arduino` series. It is built around an `ATMega` 1281 microcontroller, with 8 kB of RAM and 128 kB of programmable flash.

The `Waspmote` can be interfaced with two types of transceivers from `XBee`: one transceiver implements the `Zigbee` protocol, while the other implements IEEE 802.15.4 standard. While `ZigBee` handles node synchronization and is relatively energy efficient, it cannot be used in the present application, as it would require coordinators that have to be always on (this cannot be guaranteed in practice for a solar powered wireless sensor network). In addition, a coordinator is a single point of failure, and could potentially render a large portion of the network inoperative. We thus chose IEEE 802.15.4 `XBee` transceivers, which require the development of a routing protocol supporting multi-hop communication.

For this work, we used the standard libraries provided by `Libelium`, though we commented all unused functions from the API to increase the available memory. We also added a hard reset circuitry to enable periodic node reset, improving overall reliability.

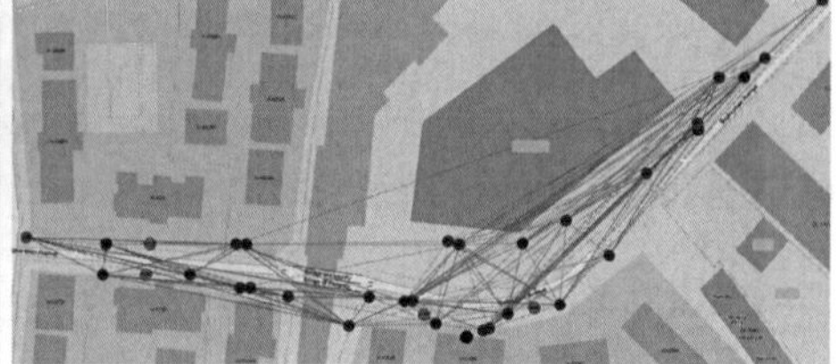

Fig. 1 Deployment of solar-powered nodes. *Left* nodes inspected before deployment. *Right* visualization of node energy

3.2 Deployment

We installed 34 nodes on both sides of a street, in an urban environment in the city of Thuwal (Jeddah, KSA). The maximal distance between any two nodes in the network is approximately 400 m, as illustrated in Fig. 1. The network has a radius of 5–6 hops, depending on the presence of parked vehicles around the main road, as well as other environmental conditions. Each node transmits its energy and local connectivity map every 300 s. The resulting data can be displayed on a visualizer built using `The Pyramid` (a `Python` web framework), and is also stored in a database.

3.3 Experimental Results

A subset of the battery voltage timeseries is illustrated in Fig. 2 below. The data collected at the sink were in voltage values between 2.7 and 4.2 V, scaled 0–100 % in the figure, each mote being represented by a different color. ADC failures can also be inferred from this figure. One month of experimental data has been used in our analysis.

4 Energy Model

4.1 Energy Generation and Storage

Each node is equipped with a rechargeable battery, connected to a solar panel through a charging chip. The complete energy management system is shown in Fig. 3 below.

The solar panel used for this study has a 3 W peak rating, which implies that it can generate at most 3 W of electrical power.

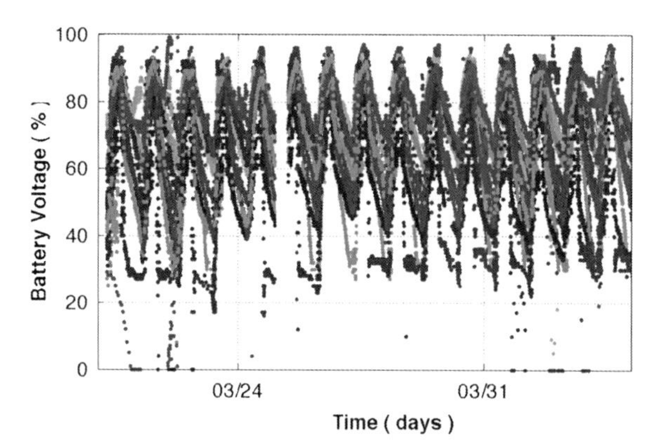

Fig. 2 Energy timeseries from 34 nodes between March 21st and April 4th, 2013. The daily energy charge-discharge cycles are clearly visible

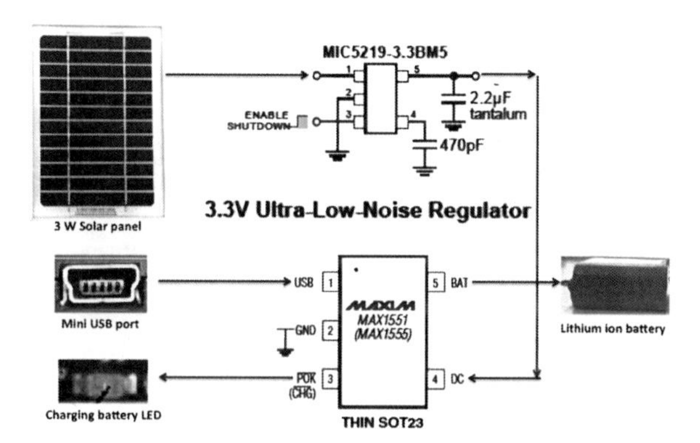

Fig. 3 Energy conversion in a solar-powered node

The charging chip efficiency ζ is the ratio of its output to input electrical power, and is around 90 % in the present case. The battery used for this study is a 2,300 mAh 1-cell Lithium Ion battery.

4.2 Energy Conservation Equation

Each node in the network sends one update regarding its energy and its RSSI and link statistics with other motes every 300 s (5 min).

All nodes are programmed with an identical software, and form a multi-hop mesh network. The average power consumption of the `XBee` transceivers in listening mode is 165 mW, while their power consumption in sending mode is 660 mW. These figures are a function of the ambient temperature, a higher temperature being associated with a higher power consumption.

Since our objective is to study the energy evolution in a wireless sensor network, we configured the sensor network in such a way that the power consumption is approximately the same for all nodes. The 300 s cycles consist in a listening phase of 150 s, during which each node sends between 1 and 34 (worst case number of messages to relay through multi-hop) messages. Thus, the total energy consumed by a node during a cycle is between 24.8 and 25.4 J (we assume that the data rate is 32 kB/s, and that each packet is 100 bytes including overhead). Since power consumptions differ by 2.5 % at most, we can assume that the overall power consumption is independent from the node, particularly since the temperature differences between nodes (caused by cast shadows) cause changes in the power consumption that are higher in magnitude. The average power draw of the transceiver is thus 85 mW, which, together with a power draw of 30 mW from the microcontroller yields a total average power consumption of 115 mW.

Let $E(k)$ be the energy stored in the node battery at (discrete) time kT. The conservation of energy [11] can be written as:

$$E(k+1) = E(k) - P_{draw}T + P_{solar}T \qquad (1)$$

where P_{draw} and P_{solar} represent the average power draw and power generated by the solar panel respectively. One of the specificities of this system is the fact that the power generated by the solar panel is not directly related to the power density of the sunlight (irradiance) received by the solar panel, because of factors related to the chain of energy conversion itself (Fig. 3). First, the solar panel and charging chip efficiencies are a function of their operating voltage, which cannot be adjusted (in particular the solar panel is not necessarily operating at its voltage of peak efficiency). Second, the power that can be actually transferred to the battery is a function of the battery condition (capacity, internal resistance) as well as its current charge (see Figs. 2 and 5). In our solar power estimation problem (see Fig. 5), our objective is to estimate the maximal available solar power, assuming that this power could be transferred to the battery.

4.3 The Need for Energy Estimation and Forecast

Despite the cyclic nature of the sun irradiance, the energy availability in a solar-powered wireless sensor network will significantly vary between nodes, for a variety of reasons:

- Large scale weather effects (uneven distribution of solar energy).
- Short scale urban shadow effects (cast shadows), these effects can appear during specific seasons (winter) depending on the latitude and the sensor network deployment locations.
- Debris, dust, humidity and salt accumulation (which are a function of the solar panel orientation).
- Battery condition and remaining capacity (which is a function of the battery history, in particular the number and the severity of deep discharges). Since a battery

that has a lower capacity is more likely to be discharged deeply during its future operation, the differences between batteries tend to get worse over time.

To enable energy aware sensing, computing and communication schemes [12], it is critical to be able to accurately forecast future (medium-term) power availability, and to estimate the current battery condition. With such parameters, the operation of the system can be scheduled in real time as a function of expected energy availability and current energy storage performance. Because of the short term (daily), medium term (seasonal) and long term (rain, dust, debris, battery and solar panel degradation) variations in energy supply, this estimation and forecast process has to be conducted relatively frequently.

Estimating the power availability in real time has additional benefits, for instance to detect faults in the battery, solar panel or charging chip. This early detection allows the optimization of the remaining energy of the node to allow enough time for its repair.

5 Estimation of Battery Condition and Capacity

5.1 Background

As all batteries, the Lithium-Ion batteries used as an energy buffer in all nodes tend to age during their operation, which translates into a loss of capacity (or as an inability to fully charge the battery) and an increase in internal resistance.

Single-cell Lithium-Ion batteries (such as the battery used in this study) have a maximal voltage of 4.2 V, and a minimal voltage of 2.7 V, below which a battery protection circuit prevents further discharge. In general, the loss of performance of a battery [13] depends on the following factors:

- Environmental conditions (in particular the temperature)
- Extended operations at full charge
- Depth, duration and number of discharge/charge cycles.

Though no battery datasheet was available to us, we used standard industry figures to evaluate the battery performance. At the beginning of this test, all batteries had been used for less than 100 charge/discharge cycles. Industry figures indicate that batteries can handle between 400–1,200 cycles before losing 30 % of their original capacity, though these values are highly dependent on environmental conditions. Therefore, our batteries can be considered as relatively new. The total duration of the experiment (3 months) added less than 100 cycles to these batteries history. The ambient temperatures measured during the experiment were comprised between 15 and 35 °C, which fall within the recommended operating temperature range for these batteries.

As stated in Sect. 4, the discharge rate during listening operations is on the order of 200 mW, while the maximal current that can be delivered by the charging chip is

280 mA, which is well within the recommended values of 1 C (2,300 mA) for charge and 2 C (4,600 mA) for discharge of standard Li-Ion batteries.

In this study, we excluded motes for which the voltage timeseries were inconsistent with the physics of the system (which denotes an ADC fault), for instance when the voltage variations between two consecutive points exceed the maximal power that can be delivered by the charging chip or the maximal power that can be dissipated by the battery. ADC failures can be directly inferred from Fig. 2.

5.2 Estimation of Battery Discharge Patterns

We first only consider the data generated during the night (i.e. after the sunset and before the sunrise), in which the solar panel generates negligible power (the illuminance caused by urban street lights is orders of magnitude smaller than the full daylight illuminance).

The battery charge decreases linearly during the night since the current drawn by the microcontroller and the peripherals is constant (approximately 35 mA). The voltage data collected during the night thus provides us information on the relationship between battery charge Q and battery voltage V. Since the loss of energy during one night does not span the complete operational range of the battery voltage, we integrate the data of multiple days to estimate the discharge curves Q(V), where Q is determined up to a constant (this does not impact our analysis however, as we are interested in the derivative of Q and not in Q itself).

We first define $t_i(V)$ as the inverse of the nighttime battery discharge timeseries $V_i(t)$ associated with nodes $i \in [1, n]$. Since the current drawn from the battery is constant,[1] we have that $Q_j(t) = Q_{0,j} - i_0 \cdot t$, or equivalently $t = \frac{Q_{0,j} - Q_j(t)}{i_0}$. Given that the $Q_{0,j}$ (initial charge) are unknown for all j, we fit these functions by solving the following least squares problem:

$$\min_{Q_{0,1},\dots,Q_{0,n}} \sum_{i,j,i \neq j} \int_0^{V_{\max}} \left(t_i(V) - t_j(V)\right)^2 dV \tag{2}$$

Problem (2) is an unconstrained quadratic programming (QP) which can be solved using standard linear algebra. We show the results of the above discharge curve fitting scheme for one node over two different time windows of 10 nights in Fig. 4. For these graphs, we assumed that 2.7 V (which is never attained in practice) corresponded to a zero charge. Note that most circuits will not discharge a lithium ion with a voltage less than 2.7 V to avoid irremediably damaging the battery.

[1] This hypothesis assumes that the current drawn by the microcontroller and its peripherals is independent of their temperature.

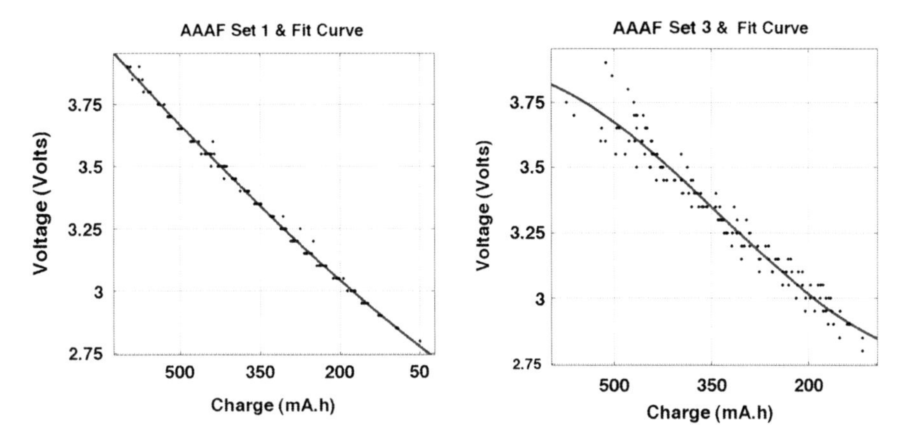

Fig. 4 Discharge *curves* of batteries AAAF evaluated on a 10 days period at the beginning of the experiment (*left*), and on a 10 days period 20 days later (*right*). The *right* subfigure shows a slight degradation in battery capacity, as well as a less reproducible (higher standard deviation) discharge pattern

5.3 Estimation of Battery Capacity

The battery capacity is estimated from the charge of the battery when its maximal voltage is reached (which is a function of the battery, see Fig. 2). The pseudo-code used for the above battery capacity estimation is presented in Algorithm 1 below. This codes relies on a polynomial fit (we used a polynomial of order 3 in practice) to estimate the charge at 2.7 V.

Algorithm 1 Pseudo-code implementation of the battery discharge curves and battery capacity estimation Algorithm

---

Input: $V_i(t)$, $t \in [t_{\text{sunset,i}}, t_{\text{sunrise,i}}]$, $i \in [1, n]$ {Voltage timeseries for i^{th} night, $i \in [1, n]$}
for $i = 0$ to $i = n$ **do**
Low pass filter(V_i)$\rightarrow$ V_i
$t_i(\cdot) := V_i^{-1}$
end for

$$\min_{Q_{0,1},\ldots,Q_{0,n}} \sum_{i,j,i\neq j} \int_0^{V_{\max}} \left(t_i(V) - t_j(V)\right)^2 dV \rightarrow Q_{0,1},\ldots,Q_{0,n}$$

Output: $(Q_{0,1}, \ldots, Q_{0,n})$ {Fitting parameters for discharge curves}
Polynomial_fit($\cup_{i\in[1,n]} Graph(Q_{0,i} + i_0 \cdot t_i(\cdot))$) $\rightarrow$ $P(\cdot)$ {Polynomial fit of the reconciliated discharge curves}
Output: Capacity $C = P^{-1}(\max_{t,i} V_i(t)) - P^{-1}(V_{\min,battery})$ {Estimated battery capacity, $V_{\min,battery} = 2.7V$ for Li-ion batteries}

As one can see from the above Table 1, the variations in capacity are significant with capacities ranging from 350 to 800 mAh. Note also that the estimation

Table 1 A systematic analysis over nine nodes yields the following estimated capacities

Mote ID	AAA8	AAAF	9CE6	AA95	0865	07EB	085A	3A02	0897
Capacity (March 1–10)	525	575	700	665	740	595	548	397	784
Capacity (March 11–20)	530	566	712	630	712	562	520	377	780
Capacity (March 21–30)	528	545	620	603	686	553	490	345	710

is fairly robust, with only minor discrepancies between battery capacity estimates. All estimated capacities are less than one third the original battery capacity, which show that Li-Ion batteries degrade quickly in outdoor environments. For our future research projects, we envision the use of Lithium Iron Phosphate batteries, which have a slightly lower energy density, but a much greater tolerance to temperature fluctuations and deep discharges.

6 Estimation of Solar Power Supply

While the battery condition and remaining capacity are a very important factor for energy management of wireless sensor networks, the solar power availability is an equally important parameter. The remaining capacity is a measure of the potential energy that can be stored in the battery, while the solar power availability can be thought of as a measure of the actual power input.

To estimate the availability of solar power, we first estimate the current generated by the solar panel during the day (similarly as in Sect. 5) from the battery voltage time series obtained during the day. The actual power input is then inferred from the product of current and battery voltage. Two examples of solar power input are shown in Fig. 5 below.

Figure 5 shows three main regimes of operation of a node:

- During the night, no solar power is generated (though the estimated generated power exhibits some noise). This corresponds to the times 0–6 and 17–24 in Fig. 5, up.
- After sunrise, some solar power begins to be generated. The solar power peaks and then reduces when the battery is close to being charged (times 6–11 in Fig. 5, up). At the beginning of this charging phase the solar power generated by the panel is close to the maximum that a solar panel can generate, provided that the battery charge is much lower than capacity, and that the generated power is not limited by the charging chip specifications.
- The power generated is then constant, equal to the power consumption of the mote when the battery is close to being charged (Fig. 5 from 11 to 17). During this phase the power that could be generated by the solar panel is greater or equal to the power consumed by the node (with equality arising around 17:00 in Fig. 5, up).

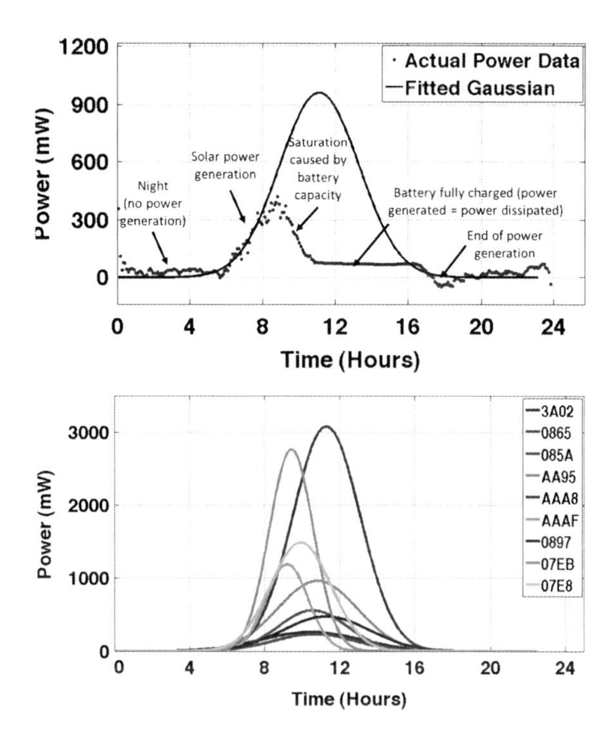

Fig. 5 Solar power estimates *Top* Actual power input inferred from Algorithm 2 on node AA95, and Gaussian fit of the estimated solar power available. *Bottom* Gaussian fits of the estimated available solar power on a subset of nine nodes, during a typical day (March 1, 2013)

Following [14], we use a Gaussian irradiance model, which has only three parameters, and can be easily fitted by the mote (this results in a three dimensional least squares problem, though simpler methods can also be used). Gaussian fit is quite accurate for our study with low complexity which require less computation at the node level. The pseudo-code used for the fitting is shown in Algorithm 2 below.

Algorithm 2 Pseudo-code implementation of the solar power estimation Algorithm

---

Input: $P(\cdot)$, $V_i(\cdot)$ {Polynomial fit of the Q(V) function, Voltage timeseries for i^{th} day}
  Low pass filter(V_i)$\rightarrow$ V_i
  Define $Q_i(\cdot) = P(V_i(\cdot))$
  Beginning and end of charge cycle detection (thresholding on Q_i') $\rightarrow$ $(t_{b,i}, t_{e,i})$
  Charge current saturation detection (zero of Q_i') $\rightarrow$ $t_{s,i}$
  Define $Pw_i(\cdot) = \frac{Q_i(\cdot)\cdot V_i(\cdot)+Pw_0}{\zeta_{\text{charging chip}}}$ {Solar panel input power}
  Gaussian_fit($Graph((Pw_{i|[t_{\min,i},t_{s,i}]\ cup[t_{e,i},t_{\max,i}]}))$ $\rightarrow$ $R(\cdot)$) {Gaussian fit of the restriction of Pw_i to the non-saturation domain}
  Output: $R(\cdot)$ {Fitting parameters for the solar power curve}

Figure 5 bottom shows Gaussian fits of the maximal solar power available to nine different nodes during March 1, 2013. As one can see from this figure, there are significant differences in solar power availability between nodes. These differences are caused by a set of factors including orientation, dust/humidity accumulation or the presence of clouds or shadows. Note that the latter factors cannot be captured by the Gaussian fit (which assumes a direct view to the sun), and would require more sophisticated models to be used.

An important assessment of the validity of the method comes from the fact that the estimated maximal power generated by the solar panel is always below its 3 W rating. Again, because of the charging chip input current limitation, not all this estimated power may be useable in practice.

7 Conclusion

This article presents a set of tools for estimating the energetic performance of solar-powered wireless sensor networks. Given the variability of solar energy availability and of the available battery capacity in a typical sensor network, we anticipate that such tools will be very useful to optimize sensing and network operations and to minimize the number of nodes that run out of energy. While ultra-low power wireless sensor network do not necessarily need such an optimization (provided that their solar panels and batteries are oversized) to run, though this suite of tools is very important for fault detection and isolation, by comparing energy and power availability with adjacent nodes.

Future work will deal with the implementation of the above methods in the nodes. All methods introduced in this article can run on typical 8 bit microcontrollers, since they mainly rely on basic thresholding and least squares fitting, which can be done efficiently with matrix operations. The required memory is on the order of hundreds of bytes, and the execution time is not expected to be an issue as computations are to be performed once daily. Also, our analysis could be extended with regard to solar power supply for the whole data for different days to conclude with similar patterns.

References

1. Ceriotti, M., Mottola, L., Picco, G.P., Murphy, A.L., Guna, S., Corra, M., Pozzi, M., Zonta, D., Zanon, P.: Monitoring heritage buildings with wireless sensor networks: the torre aquila deployment. In: Proceedings of the International Conference on Information Processing in Sensor Networks. IEEE Compute. Soc. 2009, 277–288 (2009)
2. Van Dam, T., Langendoen, K.: An adaptive energy-efficient mac protocol for wireless sensor networks. In: Proceedings of the 1st International Conference on Embedded Networked Sensor Systems, ACM, New York pp. 171–180 (2003)
3. Seada, K., Zuniga, M., Helmy, A., Krishnamachari, B.: Energy-efficient forwarding strategies for geographic routing in lossy wireless sensor networks. In: Proceedings of the 2nd International Conference on Embedded Networked Sensor Systems, ACM, New York pp. 108–121 (2004)
4. Huang, H., Hu, G., Yu, F.: Energy-aware geographic routing in wireless sensor networks with anchor nodes. Int. J Commun. Syst. **26**(1), 100–113 (2013)

5. Alippi, C., Anastasi, G., Di Francesco, M., Roveri, M.: Energy management in wireless sensor networks with energy-hungry sensors. IEEE Instrum. Meas. Mag. **12**(2), 16–23 (2009)
6. Feeney, L.M., Andersson, L., Lindgren, A., Starborg, S., Ahlberg Tidblad, A.: A testbed for measuring battery discharge behavior. In: Proceedings of the 7th ACM International Workshop on Wireless Network Testbeds, Experimental Evaluation and Characterization, ACM, New York pp. 91–92 (2012)
7. Fonseca, R., Dutta, P., Levis, P., Stoica, I.: Quanto: Tracking energy in networked embedded systems. In: OSDI, vol. 8, pp. 323–338 (2008)
8. Dunkels, A., Osterlind, F., Tsiftes, N., He, Z.: Software-based on-line energy estimation for sensor nodes. In: Proceedings of the 4th Workshop on Embedded Networked Sensors, ACM, New York pp. 28–32 (2007)
9. Saha, B., Goebel, K.: Modeling li-ion battery capacity depletion in a particle filtering framework. In: Proceedings of the Annual Conference of the Prognostics and Health Management Society (2009)
10. Park, C., Lahiri, K., Raghunathan, A.: Battery discharge characteristics of wireless sensor nodes: an experimental analysis. Power **20**, 21 (2005)
11. Raghunathan, V., Kansal, A., Hsu, J., Friedman, J., Srivastava, M.: Design considerations for solar energy harvesting wireless embedded systems. In: Proceedings of the 4th International Symposium on Information Processing in Sensor Networks, IEEE Press, 64 (2005)
12. Bouabdallah, F., Bouabdallah, N., Boutaba, R.: On balancing energy consumption in wireless sensor networks. IEEE Trans. Veh. Technol. **58**(6), 2909–2924 (2009)
13. Sikha, G., White, R.E., Popov, B.N.: A mathematical model for a lithium-ion battery/electrochemical capacitor hybrid system. J Electrochem. Soc. **152**(8), A1682–A1693 (2005)
14. Tsai, H.L.: Insolation-oriented model of photovoltaic module using matlab/simulink. Solar Energy **84**(7), 1318–1326 (2010)

Experiences with Sensors for Energy Efficiency in Commercial Buildings

Branislav Kusy, Rajib Rana, Phil Valencia, Raja Jurdak and Josh Wall

Abstract Buildings are amongst the largest consumers of electrical energy in developed countries. Building efficiency can be improved by adapting building systems to a change in the environment or user context. Appropriate action, however, can only be taken if the building control system has access to reliable real-time data. Sensors providing this data need to be ubiquitous, accurate, have low maintenance cost, and should not violate privacy of building occupants. We conducted a 3 year study in a mid-size office space with 15 offices and 25 people. Specifically, we focused on sensing modalities that can help improve energy efficiency of buildings. We have deployed 25 indoor climate sensor nodes and 41 wireless power meters, submetered 12 electric loads in circuit breaker boxes, logged data from our building control system and tracked activity on 40 desktop computers. We summarize our experiences with the cost, data yields, and user privacy concerns of the different sensing modalities and evaluate their accuracy using ground-truth experiments.

1 Introduction

Climate change and greenhouse gas emissions are forcing society to reevaluate energy efficiency practices. Buildings consume a large portion of the total energy in industrial countries and most of this energy is provided by conventional fossil-based power plants. As a result, numerous techniques for improving energy efficiency of commercial buildings have been proposed in recent years. Examples include lighting systems and window shaders that react to changing daylight and occupancy conditions [1], software products that manage and control power consumed by PC

B. Kusy (✉) · R. Rana · P. Valencia · R. Jurdak
CSIRO, Autonomous Systems, Brisbane, QLD, Australia
e-mail: brano.kusy@csiro.au

J. Wall
CSIRO, Energy Technology, Newcastle, NSW, Australia

K. Langendoen et al. (eds.), *Real-World Wireless Sensor Networks*,
Lecture Notes in Electrical Engineering 281, DOI: 10.1007/978-3-319-03071-5_23,

workstations [2], and Heating Ventilation and Air Conditioning (HVAC) systems that adapt to occupancy, electricity price, and weather forecast [3–5].

Many of these systems rely on real-time sensing of environmental data and user context. Future building control systems will no doubt deploy and integrate numerous sensors in their core functionality. The advantage of the sensor and actuator convergence in building control systems is that individual sensors can be reused for different tasks. For example, power-meters can be used to detect occupancy of offices based on power consumption of desktop PCs and LCD monitors [6]. Thermal comfort of people can be estimated using office temperature and user feedback [7]. The overhead of installation and maintenance of building sensors is thus expected to be amortized by the costs of operating the whole building. Despite the exploration of multiple sensing modalities for promoting energy efficiency in buildings, there is still limited work on comprehensive comparison of their effectiveness.

The goal of this work is to quantify advantages and disadvantages of different sensing modalities in commercial office buildings. We collected data over a 3 year period from environmental sensor nodes with temperature, light, humidity, and presence sensors deployed in every office in our building and from power-meters measuring energy usage of office appliances. We also contracted electricians to install submeters on load circuits and obtained access to data from our building control system. Finally, we developed a PC application that tracked activity of the users on desktop computers and allowed us to remotely survey users about their thermal comfort.

We evaluate installation and operational costs of the different building sensors. We found that data yields of low cost sensors tend to decrease over time due to hardware failures. Battery operated devices introduce frequent black-out periods due to lack or negligence of maintenance. Software-based sensing solutions are also surprisingly difficult to maintain, despite existing IT policies of pre-installing the software on new computers. Permanently powered commercial products, on the other hand, perform reliably on a long-term basis.

We also evaluate accuracy of the different sensors by conducting ground truth experiments. We mostly focus on detecting office occupancy and estimation of thermal comfort of individual users. Our findings show that power-meters are 10–20 % less accurate in detecting office occupancy than passive infrared sensors and suffer from high false positive rates. PC activity loggers perform better than power-meters, but are generally rejected by users due to privacy concerns. Interestingly, users accept power-meters connected to their desktop PCs despite us disclosing their use for occupancy detection.

Finally, we conducted several human-in-the-loop experiments aiming at energy usage reduction of HVAC and office appliances. Our empirically-based simulations indicate that 42 % reduction in HVAC energy usage can be achieved over a centrally managed HVAC with static space temperature setpoints. We also found that presenting energy efficiency data to users can lead to behavior change and energy reduction, achieving 10 % energy savings on office appliances. Our findings help justify the need for fine-grained continuous sensing of office occupancy, office climate, and energy usage for building-scale systems.

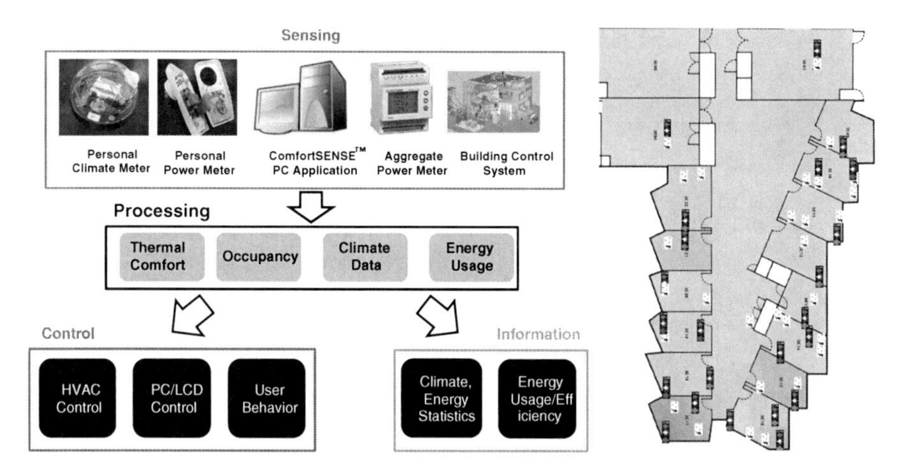

Fig. 1 *Left* Data from low level sensors can be used to drive adaptive environment control and improve our understanding of the building energy usage. *Right* Office building used for our deployment. Icons depict locations of sensors in offices

2 Experimental Setting

We built a comprehensive system for collecting data relevant to energy efficiency and energy usage in commercial buildings. Specifically, we focused on collecting information about (1) presence of people in their offices, (2) energy usage of office appliances and the HVAC, and (3) subjective perception of thermal comfort of users. Figure 1 (left) provides an overview of our architecture that includes multiple sensor inputs within a typical office building. Real-time collection of this data is important for closing the loop between building control, user behavior and comfort, and changes in the environment.

We deployed our system on a single floor of an office building that included three single-occupancy offices, 10 double-occupancy offices, one conference room, and two open-plan student offices (see Fig. 1 right). We describe each of our five sensing techniques in more detail in the following text.

Personal climate monitors (climate) and power meters (power) We built two sensor platforms, climate meters and power meters, based on our in-house sensor network board. The board uses Intel 8051 processor and sub 1-GHz Nordic radio and is optimized for low power and small size. The nodes self-organize to provide a multi-hop wireless mesh and run on a single battery charge for approximately 2 months. We sampled climate and power meters every 60 s and transmitted the data over the wireless network to a central database. The climate monitors also transmitted their current battery voltage level for maintenance purposes.

We deployed one *climate monitor* per user, most frequently next to the user's LCD monitor. We measured office climate and occupancy using temperature, humidity, light, and motion sensors.

The *power meters* measured current, voltage, power, and cumulative energy consumption of electric appliances at a power socket level (similar to [6]). We augmented off-the-shelf power meters with our wireless sensor board to enable real-time wireless data collection of the energy data. We were limited to two power meters per user, thus we generally measured an aggregate consumption of a desktop PC and LCD monitor with one power meter and used the other power meter for external hard drives, laptops, and fans.

Commercial power meters (BCM) We installed Powerlogic PM9C power meters in breaker circuits (thus we call this modality BCM) to measure the aggregate energy usage of 12 electric circuits in our building, including user office sockets, lighting, hot-water heaters, and printers. We connected the meters to our internal Ethernet network and periodically recorded voltage, current, and power on each of the electrical circuits every 60 s in the database.

Building control and management system (BMS) Our building is controlled by Tridium JACE controllers connected together by an internal modbus network. The access to the controller is restricted to our property managers who use it on a daily basis, primarily to inspect and fine-tune performance of the heating, ventilation, and air-conditioning (HVAC) system. We have obtained access to the system and used standard Open Building Information Exchange (oBIX) RESTful web interface to log data in the database and to control temperature in the offices during user trials. Our office space is logically divided into four air-conditioning zones and we used a python script to collect 11 sensor inputs per air-conditioning zone, including information about chillers, heaters, supply fans, indoor temperature, and space temperature setpoint that the HVAC is currently set to. Unlike the fine-grained office-level temperature data obtained by the personal climate monitors, BMS has one temperature sensor per zone.

PC application (CSENSE, SURVEY) Finally, we used *ComfortSENSE*™ PC application [8] to track user activity on desktop computers and to administer user surveys. The application runs as a background process on desktop computers and can be controlled from a central server to display user surveys on all registered computers. Surveys focus mainly on user thermal comfort and their satisfaction with HVAC performance. Additionally, the application tracks key-strokes and mouse events on the host operating system and records this information every 5 min in the database. Users can opt-out of activity monitoring in case of privacy concerns.

3 Analysis of Empirical Data

We summarize data that we collected over the course of our study in Fig. 2. The rows in the figure denote different sensing modalities (see Sect. 2). The columns show the total amount of data received from all nodes for a given modality (top figure) and the total number of nodes that sent at least one data point (bottom figure) for a given month. For clarity, we normalize each row by the maximum value found for that modality over the whole experiment. The normalization factors, shown as

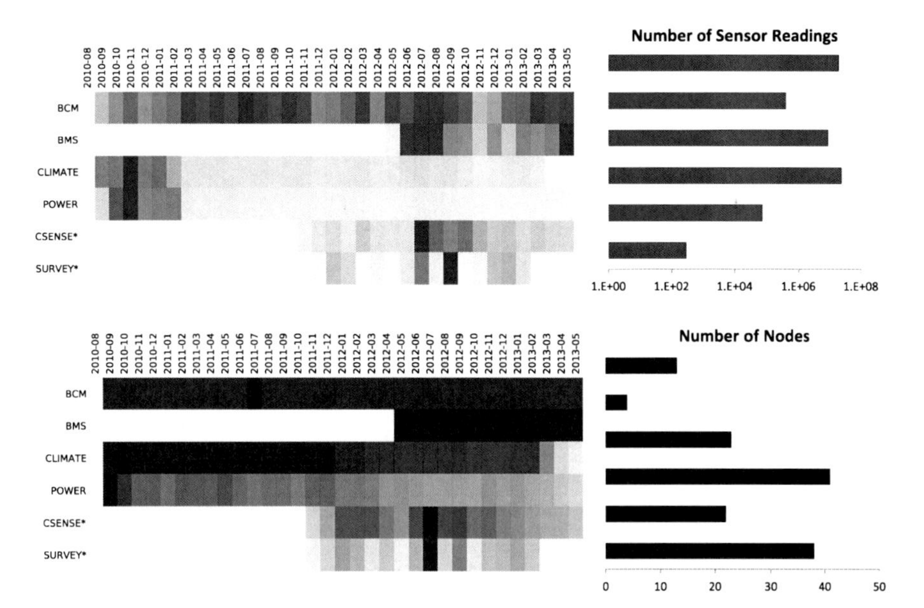

Fig. 2 Summary of data that we downloaded over 3 years for all sensor modalities. Figures on the *left* show number of data points for each month, normalized to the maximum value over all months for each row. Figures on the *right* show the normalization factor (i.e., the maximum value over all months). *Note* the sensor readings are in a logarithmic scale, while the number of nodes are in linear scale

bar-graphs in Fig. 2, vary for each modality, as the number of sensor nodes and sampling rates are different. For example, the dark blue color corresponds to 20 million and 290 sensor readings for the BCM and SURVEY* modalities, respectively.

We started our experiment by deploying climate, power, and BCM meters in September 2010. Overall, 23 climate domes, 41 power meters, and 12 BCM meters were deployed. We started experimenting with HVAC and ComfortSENSE™ in November 2011 and started logging BMS data in May 2012.

Installation and maintenance cost By far the most expensive was the initial deployment period. BCM meters cost us approximately \$12k, including the cost of professional installation in our building. Manufacture and deployment of the wireless sensor platforms cost approximately \$12k, or about \$200 per node, including batteries and enclosures. Clearly, software-based solutions or solutions that utilize existing infrastructure are advantageous over new sensors. We obtained the BMS, and CSENSE modality for free, as they are amortized by the costs of running the whole building. In terms of the on-going maintenance, the only significant cost in terms of dollars was replacing the rechargeable batteries in CLIMATE nodes after one and half years.

While we spent minimal time maintaining BCM and BMS modalities, maintaining CLIMATE, POWER, and CSENSE modalities was more difficult. CLIMATE and

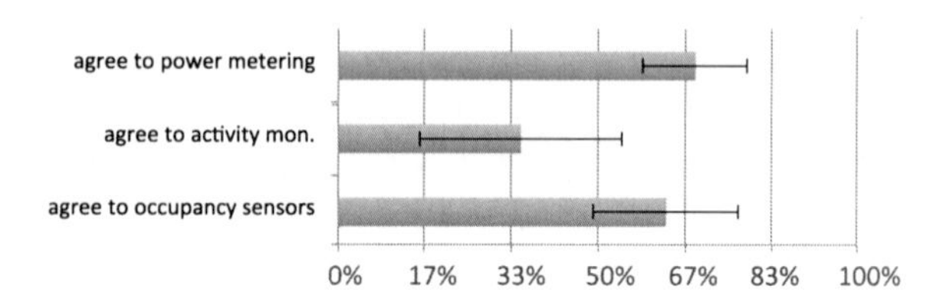

Fig. 3 Survey of six participants on the intrusiveness of different sensing modalities

POWER nodes break down frequently as sensors or connectors fail and people tend to uninstall ComfortSENSE™ application after a few months.

Data yields It is clear that the best performing sensors in terms of node failures are BCM and BMS. This is commercially tested hardware that is periodically maintained by our property managers. The lower yields in the sensor reading graphs tend to happen due to power outages on the server, for example, around Christmas when the building effectively shuts down.

All other sensing modalities tend to have decreasing yields over time. For example, the number of functional POWER meters dropped from 41 to less than 20 after 1.5 years to less than 10 at present. CLIMATE nodes have frequent outages due to flat batteries. COMFORT logging tends the perform best when we run targeted HVAC experiments as shown by the spikes on the SURVEY modality. This is because we tend to walk around and manually check that everyone is running ComfortSENSE™ application. Surprisingly, the high failure of ComfortSENSE™ was observed despite a company-wide IT policy that automatically installs the application on all new computers.

User privacy Finally, we surveyed our users to evaluate their privacy concerns for each sensing modality. Figure 3 shows that a large number of people reject computer-based data loggers. We disabled PC activity sensing for these users for the duration of the experiment. Interestingly, users were not as sensitive to the indirect occupancy sensing through power meters, despite knowing that the power data information is used to determine occupancy.

4 Occupancy Estimation

Occupancy estimation offers energy savings by fully or partially shutting off HVAC in unoccupied offices. We developed three threshold-based algorithms to evaluate the accuracy of office occupancy sensing based on three sensing modalities—PIR sensors (collected from climate monitors), power meter readings (collected from power meters), and ComfortSENSE™ computer activity detection. We determined thresholds for distinguishing presence and absence of a person in an office by running ground truth experiments using a web-cam. We tracked occupancy of 4 offices over multiple days and split this data randomly to training and testing sets. The training set is to find thresholds for office occupancy and the testing set is used to evaluate accuracy of the occupancy detection.

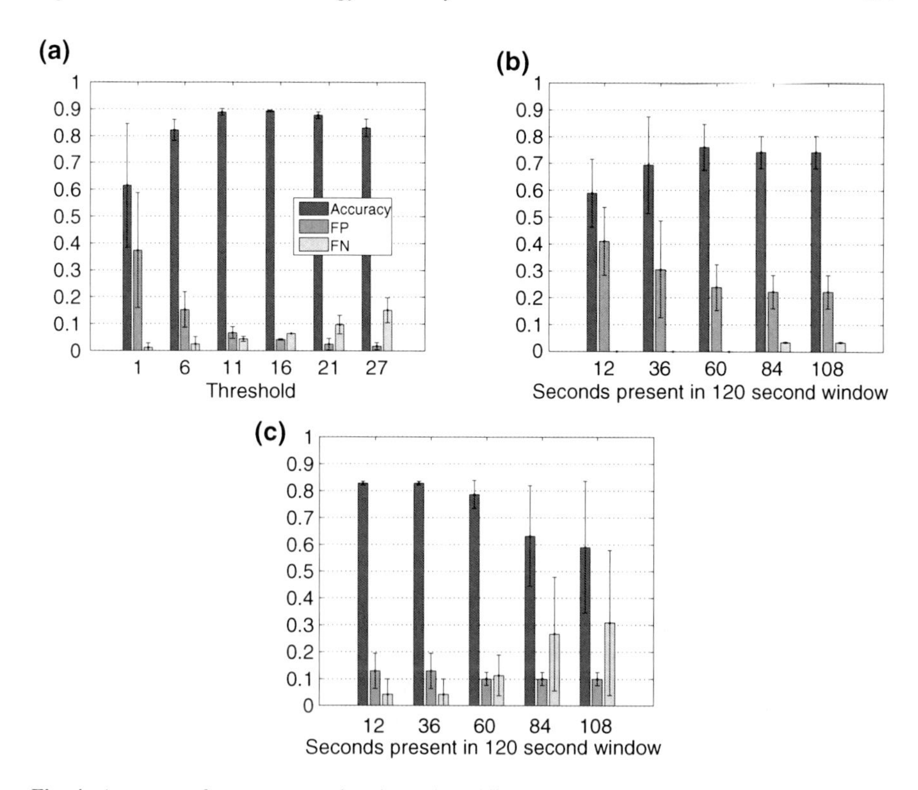

Fig. 4 Accuracy of occupancy estimation using different sensing modalities. FP and FN refer to false postive and false negatives respectively. **a** PIR sensors. **b** Power meters. **c** ComfortSENSE™

Based on our investigations, we found that the threshold for PIR sensor readings is generic and one single threshold fits all people and offices. However, determining a generic threshold for power meter was not achievable. This is because, different people have different devices (even the number of monitors vary) connected to the power meters. Another issue was the different screen saver intervals that people use with their LCD monitors. For example, if a person uses a 10 min interval, the power meter current will only drop 10 min after the person left the office. Therefore, we had to find user-specific thresholds for power meter readings to distinguish the occupied and empty state for each user.

Figure 4 shows the results. PIR sensing outperforms both the power meters and the PC application, but the difference is less than we expected. PIR-based algorithm reaches its optimum at a threshold, where short periods of inactivity do not yet trigger false negatives. Power meters suffer from significant number of false positives as the screen-saver thresholds are set to a relatively long period on average and the algorithm fails to detect that people have left the office. PC activity threshold is a tunable parameter in the ComfortSENSE™ application, which gives us a more fine-grained control over false positives and true negatives.

Occupancy detection has been studied previously with similar results. For example, PIR sensors can achieve 96 % accuracy of office occupancy detection when combined with a reed switch [9]. Camera systems can detect the number of people in an office with 80 % accuracy [5]. Our focus was on sensing modalities that are likely to be deployed in buildings at low cost and that do not compromise user privacy. We show that simple data thresholding compares favorably to previous results and plan to further improve its performance by fusing data across all available sensing modalities.

5 Thermal Comfort

The American Society of Heating, Refrigerating and Air conditioning Engineers (ASHRAE) defines Thermal comfort as "the state of mind in humans that expresses satisfaction with the surrounding environment" [10]. It is rated using the ASHRAE Thermal Sensation Scale [10], which has the following definition: 3 = hot; 2 = warm; 1 = slightly warm; 0 = neutral; -1 = slightly cool; -2 = cool; -3 = cold. Knowing the thermal comfort of office occupants can help saving energy on cooling or heating. Specifically, thermal comfort models can give us lower and upper bounds on the temperature at which a majority of occupants are comfortable.

We collected 849 comfort surveys from our office occupants, which were mapped to the temperature collected from a climate dome associated with each individual. Figure 5 (left) shows temperature versus comfort mapping. We observe that for a wide range of temperatures (16.5–21 degree), thermal comfort undergoes a very small change: -0.5 to 0.5. A change of one degree centigrade of temperature can save approximately 15 % of energy [11], therefore there is a tremendous potential of saving energy if people could tolerate slight discomfort.

We demonstrate that it is possible to build accurate models of user thermal comfort provided accurate temperature sensing and the corresponding comfort surveys are available. Due to the sparsity of the surveys, it is not guaranteed that we will have a comfort level for all possible temperature readings. We used ϵ-support vector regression(ϵ-SVR) [12] to build the prediction model given the temperature input. For various comfort ranges, we divide the comfort surveys into training and test set and conduct predictions. We observe that prediction accuracy is over 91 % for all possible range of comfort levels (see Fig. 5 right).

Note that in order to find a generalized temperature threshold, we need data across different climates and buildings as was investigated in [13], which is out of the scope of the chapter. In this chapter we mainly demonstrate that given a sparse data of temperatures and corresponding thermal comfort responses, we can accurately predict thermal comfort for unknown temperatures.

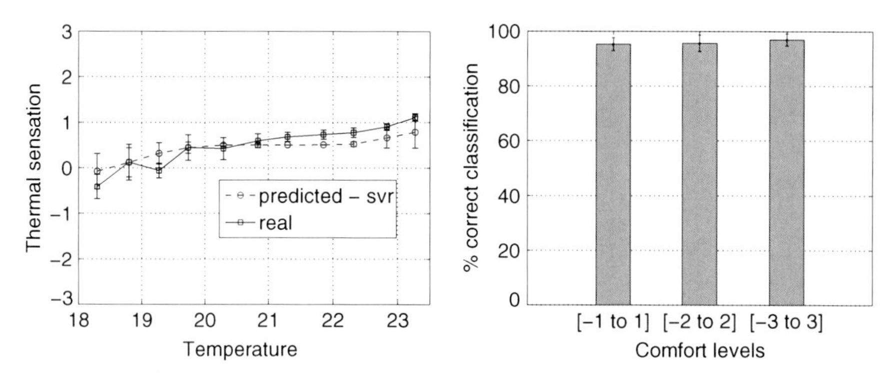

Fig. 5 *Left* Temperature versus thermal comfort. *Right* SVR prediction accuracy

6 Energy Savings

One of our main goals is to show that building sensors can help to improve energy efficiency in commercial buildings. We describe two experiments that confirm viability of our vision in this section.

6.1 Office Appliances

Our study was limited to office spaces that have appliances, such as PCs, monitors, printers, and audio-visual equipment. However, similar ideas can be readily applied to laboratory, kitchen, and other equipment connected to the grid.

We explored two strategies to reduce energy consumption. First, we identified opportunities for energy savings by quantifying the energy cost of user actions. We defined energy usage efficiency through occupancy and considered any energy used in empty offices as wasted. We tracked individual performance on a regular basis, measured wasted energy, observed energy usage trends, and identified the maximum energy users in the lab. Some of these results were distributed within our lab and discussed at our group meetings.

Second, we helped people save energy in an automated way. Users had access to our sensor data and could use our scripts to automatically improve energy efficiency of their PCs and LCDs. For example, users could reset their LCD monitor sleep timeout when away from the computer. Despite lacking a comprehensive analysis at the moment, we saw first indications of user involvement in sustainable practices when provided with energy accountability tools.

We first established a baseline power consumption estimates for 15 participants over a period of 5 weeks. During this period, participants were informed that energy consumption was being measured, however they were provided with no feedback on their actual consumption.

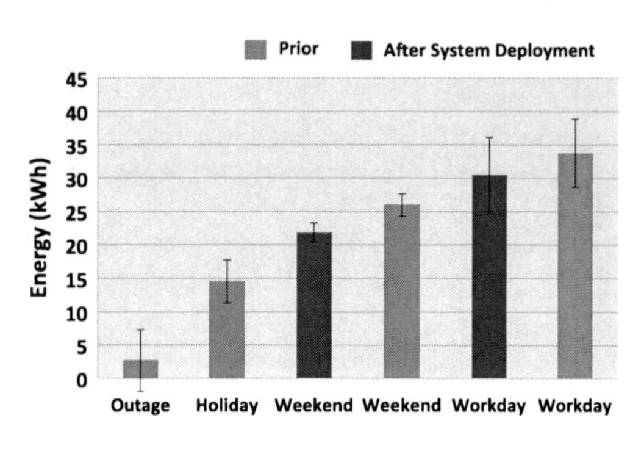

Fig. 6 Per-day energy usage on workdays and during weekends aggregated over 15 offices, two printers, and a conference room

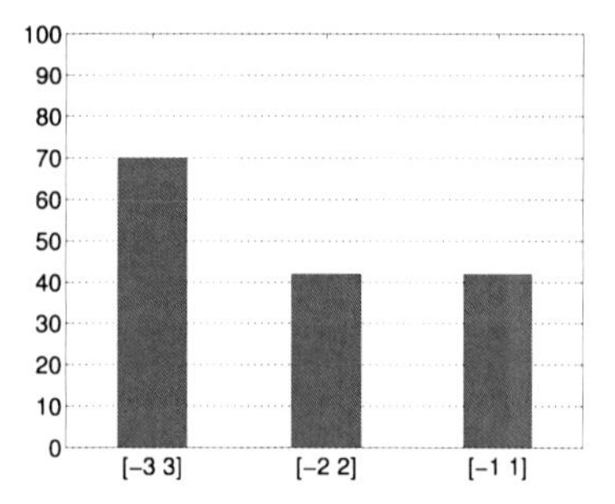

Fig. 7 Energy savings over non-adaptive HVAC, using different definitions of comfort

Over the subsequent 2 weeks we let participants see their personal energy usage and energy efficiency trends. This fostered an energy-aware culture which encouraged participants to turn off monitors, computers and other mains-powered devices where possible. Figure 6 illustrates results of this study. We average daily energy consumption over the whole lab and report the values measured during workdays, weekends, and holidays. The difference between holidays and weekends is that some students come to work on the weekend, while nobody was present at work during holidays. We consider the holiday power consumption as the lower bound for the power consumption, as only servers of some of the employees are left running. We also report energy used during building-wide power outage to illustrate the energy used by the critical infrastructure protected by UPS. Overall, our system managed to reduce energy consumption by 10 % on workdays and 16 % on the weekends.

Systems targeting behavioral change suffer from the *user fatigue* problem where majority of users slip back to their old habits as the initial excitement fades [14]. We suspect our system will face similar problems in real-world deployments. The mechanisms for automating energy savings will, however, alleviate user fatigue and we intend to develop this area further in our future work.

6.2 HVAC

Empirical testing of HVAC in real-world office buildings is often impractical. We, therefore, provide analysis of how much savings we can expect in a typical office building. We propose to save energy by two key actions: (1) choosing optimal temperature that minimizes the difference between room temperature and outside temperature subject to satisfaction of occupants thermal comfort; and (2) turning off the HVAC when a zone is unoccupied. We built a simple Matlab simulator that replays empirical occupancy data and allows for different HVAC control strategies. We compare the current strategy used in our building (fixed space temperature setpoint) to a system that dynamically computes optimal temperature based on the comfort models of occupants.

We compute the energy savings over 3 months period: March 2011 to June 2011, where March to May is in Autumn and June is in winter. We choose the time interval of recalculating the setpoint to be relatively large: 30 min, since changing the HVAC settings every minute could be prohibitively expensive. Furthermore, in order for the "set temperature" to take effect it may require at least 15 min [9]. The temperature setpoint for the fixed strategy was chosen a conservative 25 °C. As we mentioned before, prior work has shown that office buildings can save as much as 15 % energy for each 1 °C increase in setpoint temperature in the summer [11].

In order to calculate energy savings, we first compute the difference between the optimal setpoint and the fixed setpoint at any point in simulated time. We then calculate the sum of all temperature differences over the course of the experiment and then convert the aggregate temperature difference to energy savings by multiplying by 15. Average energy savings while maintaining various comfort levels are summarized in Fig. 7.

We observe that energy savings for the comfort level [−3 to 3] (i.e., −2 to −3 and 2 to 3) is very high (nearly 70 %). This is because −3 and 3 are two extreme bounds, therefore, the HVAC system needs to perform minimum or no work. Energy savings for other ranges are still quite large (approximately 42 %). One interesting point to note is that energy savings for ranges [−2 to 2] and [−1 to 1] are similar, therefore, choosing better comfort level [−1 to 1], we can still save substantial amount of energy.

7 Related Work

Jung et al. [15] attempt to determine the minimum number of power meters adaptively to validate and improve the appliance state (ON/OFF) estimation on a centralized home energy monitor. Several recent works have investigated the use of occupancy sensing to reduce building energy consumption. Newsham and Birt [16] use an autoregressive integrated moving average model (ARIMA) to predict the demand for energy in a building based on occupancy sensor data, which includes passive infrared

(PIR), door, and carbon-dioxide sensors. Hsu et al. [17] introduce an information architecture for energy usage in CyberPhysical Systems (CPS), which includes a smart phone, RESTful web services, and low power sensors (tags). Agrawal et al. evaluate occupancy sensors to drive a joint HVAC and appliance control system in a shared office environment for the purpose of building energy management [9]. Their work uses a combination of PIR and office door state (magnetic reed) sensors to reduce false alarms. While our work also combines HVAC and appliance measurement and control, it additionally tracks thermal comfort levels through continuous user feedback and jointly considers comfort as part of the overall building control system.

Ploennigs et al. [18] attempt to measure thermal comfort by introducing the concept of a virtual sensor based on simple sensors, such as temperature. They use artificial neural networks to model comfort levels based on simple sensor readings. The work in [5] combines the concept of comfort and HVAC control. They collect and use building occupancy data to achieve up to 40 % in building energy savings subject to COMFORT standards. They use low cost image sensors to detect occupancy state transitions, and Markov chains to estimate building occupancy. While our work shares the goal of maintaining thermal comfort levels while saving energy in buildings, our work uses a combination of PIR and current sensor readings to determine occupancy, which yields higher accuracy. In addition, our system combines the HVAC energy control, based on Wall et al. work [19], with appliance energy savings through individual power meters and feedback back to users. Finally, our system is a closed-loop system that enables users to affect their thermal comfort through a visual interface, which allows for dynamic optimization of the comfort/HVAC energy tradeoff.

8 Conclusions

In this chapter, we described our experience with a long-term comprehensive study of building sensors for energy usage and energy efficiency, comprised of wireless pervasive sensors, building control units, and user activity detection software. We evaluate the long-term cost, data yields, and user acceptance of different sensing modalities and evaluated sensing accuracy using algorithms for estimation of building occupancy and thermal comfort of individuals. Using this higher level information, we can estimate that an adaptive HVAC system can save up to 42 % of total energy and energy efficiency of individual employees can lead to 10 % reduction of energy used by office appliances.

References

1. Singhvi, V., Krause, A., Guestrin, C., Garrett, Jr., J.H., Matthews, H.S.: Intelligent light control using sensor networks. In: SenSys '05, ACM, pp. 218–229. New York (2005)
2. GreenTrac. http://www.greentrac.com/
3. Agarwal, Y., Balaji, B., Gupta, R., Lyles, J., Wei, M., Weng, T.: Occupancy-driven energy management for smart building automation. In: BuildSys '10, ACM, pp. 1–6. New York (2010)
4. BuildingIQ. http://www.buildingiq.com
5. Erickson, V., Carreira-Perpinan, M., Cerpa, A.: Observe: occupancy-based system for efficient reduction of HVAC energy. In: IPSN'11, pp. 258–269 (April 2011)
6. Jiang, X., Dawson-Haggerty, S., Dutta, P., Culler, D.: Design and implementation of a high-fidelity ac metering, network. In: IPSN'09, pp. 253–264 (2009)
7. Erickson, V.L., Cerpa, A.E.: Thermovote: participatory sensing for efficient building hvac conditioning. In: BuildSys '12, ACM, pp. 9–16. New York (2012)
8. Wall, J., Ward, J., West, S., Piette, M.: Comfort, cost and co2-intelligent hvac control for harmonising hvac operating principles. In: IIR HVAC Energy Efficiency Best Practice Conference, Melbourne (2008)
9. Agarwal, Y., Balaji, B., Dutta, S., Gupta, R., Weng, T.: Duty-cycling buildings aggressively: the next frontier in HVAC control. In: IPSN'11, pp. 246–257 (2011)
10. ANSI/ASHRAE Standard 55. Thermal Environmental Conditions for Human Occupancy (2010)
11. Ward, J., White, S.: Smart thermostats trial. Tech. rep. CSIRO (2007)
12. Vapnik, V.N.: The Nature of Statistical Learning Theory. Springer, New York (1995)
13. Rana, R., Kusy, B., Jurdak, R., Wall, J., Hu, W.: Feasibility analysis of using humidex as an indoor thermal comfort predictor. Energy Build. (2013)
14. Jiang, X., Van Ly, M., Taneja, J., Dutta, P., Culler, D.: Experiences with a high-fidelity wireless building energy auditing network. In: SenSys '09, ACM, pp. 113–126. New York (2009)
15. Jung, D., Savvides, A.: Estimating building consumption breakdowns using on/off state sensing and incremental sub-meter deployment. In: SenSys '10, ACM, pp. 225–238. New York (2010)
16. Newsham, G.R., Birt, B.J.: Building-level occupancy data to improve arima-based electricity use forecasts. In: BuildSys '10, ACM, pp. 13–18. New York (2010)
17. Hsu, J., Mohan, P., Jiang, X., Ortiz, J., Shankar, S., Dawson-Haggerty, S., Culler, D.: Hbci: human-building-computer interaction. In: BuildSys '10, ACM, pp. 55–60. New York (2010)
18. Ploennigs, J., Hensel, B., Kabitzsch, K.: Wireless, collaborative virtual sensors for thermal comfort. In: BuildSys '10, ACM, pp. 79–84. New York (2010)
19. Wall, J., Ward, J.K., West, S.: Autonomous controllers for intelligent hvac management. In: World Sustainable Building Conference, pp. 1259–1267 (2008)

Wireless Sensor Networks for Building Monitoring Deployment Challenges, Tools and Experience

Alan McGibney, Suzanne Lesecq, Claire Guyon-Gardeux, Safietou R. Thior, Davide Pusceddu, Laurent-Frederic Ducreux, François Pacull and Dirk Pesch

Abstract As more large scale deployments of wireless solutions come on stream there is a real need for deployment support for system integrators to ensure a reliable infrastructure can be achieved. This chapter presents the experiences gained from a real deployment and discusses the process used by a designer in order to investigate what value the use of current support tools can offer a designer.

1 Introduction

In the building automation space, there is a perception of poor reliability and unpredictability when wireless is used over its wired counterparts. However, particularly in retro-fit scenarios, the benefits of wireless technology far out weighs this risk. It is proposed that the lack of confidence in wireless solutions for buildings is borne out due to the lack of formal design and deployment methodologies that can be followed by even inexperienced designers to ensure a reliable communication infrastructure can be achieved. There has been a number of research works focusing on deployment issues highlighting typical problems with WSN such as coverage, link or node problems [1–3]. This chapter will focus on the deployment process and will present the challenges and experience of a real-world deployment in a building following currently available industry guidelines. This approach will then be compared against a methodology using deployment planning support tools to aid in the definition of device placement within the specific environment. The advantage of using tool support over the more traditional ad-hoc deployment process will be investigated.

A. McGibney (✉) · D. Pusceddu · D. Pesch
Cork Institute of Technology, Cork, Ireland
e-mail: alan.mcgibney@cit.ie

S. Lesecq · C. Guyon-Gardeux · S. R. Thior · L.- F. Ducreux · F. Pacull
CEA, LETI Minatec Campus, Grenoble, Fance

K. Langendoen et al. (eds.), *Real-World Wireless Sensor Networks*,
Lecture Notes in Electrical Engineering 281, DOI: 10.1007/978-3-319-03071-5_24,

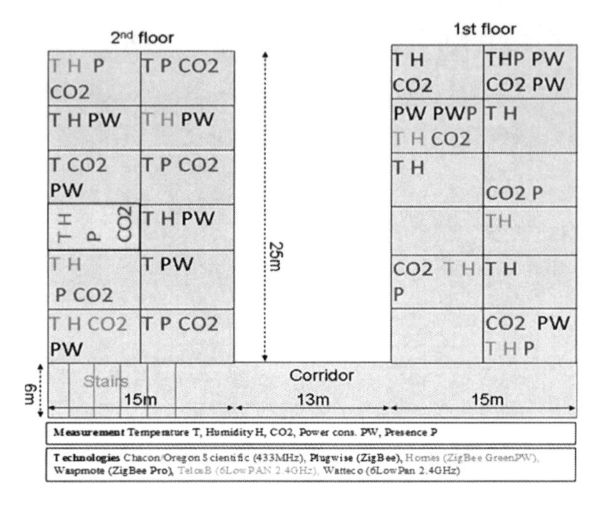

Fig. 1 Sensor distribution in the U-shape office building

2 Problem Specification

A three storey office building with a footprint of $1600\,\mathrm{m}^2$ was used for the evaluation. The building is constructed of reinforced concrete with an external thermal insulation made of bricks secured on metallic rails fixed on the outside of the walls. A number of devices were made available by the building occupier that were reused to minimize the overall cost. This led to a heterogeneous deployment of different wireless technologies with RF communication links in the frequency bands of 433 MHz, 868 MHz and 2.4 GHz. Devices have been spatially distributed in a non-uniform way with each technology spread over both wings, see Fig. 1. Data must be collected on a periodic bases, preferably on a unique computer located in one office (highlighted on the map) on the 2nd floor. The power consumption sensors (PW) and CO_2 level sensors are line powered and must be located at wall plugs. Two sensors with the same color (e.g. temperature T and relative humidity H in the first office on the 1st floor) must be co-located. Two deployment approaches will be considered for this site. Firstly, a Baseline deployment is undertaken which is based on a designer's experience along with any available manufacturer guidelines for each technology. An iterative approach is taken for each technology to define the number and position of gateways and repeaters when necessary. To evaluate if planning tools can improve on current approaches, a second approach known as Design Tool deployment was undertaken. This involved using a software package to model connectivity and generate the deployment plans for each technology.

3 Baseline Deployment

The technologies chosen present a large variety of topologies, communication ranges, leading to different deployment challenges. Note that a unique data collection point is preferred in order not to spread gateways all over the building. Thus, due to the limited indoor range for some devices (e.g. 5–10 m for ZigBee), repeaters are expected to be mandatory in order to ensure a high delivery ratio. Also several WiFi access points exist in the building, which will add extra communication difficulties. Devices on the 433 MHz frequency band implement proprietary protocols. From expert know-how, their indoor communication range is around 20 m. The topology must be a star, which might result in the use of extra gateways to collect data. Devices on the 868 MHz frequency band have also been used. When compared with the 2.4 GHz devices, the indoor communication range seems better, with 20–30 m (manufacturer information).

There are little support tools available for the technologies used in the building. Therefore, range planning guidelines [4] are used, where the communication range for each sensor of a particular technology is assumed constant and the signal coverage is idealized based on a circle. As part of the guidelines, it is usually recommended that for a robust network, redundant radio receiver paths should be implemented through the inclusion of additional repeaters. The guidelines also endorse that the deployment plan should be verified and adjusted based on a site survey through the addition/removal of repeaters or gateways. Note that this verification was done during the installation phase. Maps of the building were prepared in svg format with the specified sensing locations that were defined based on the initial requirements and deployment constraints. Circles (representing the communication range) drawn around these devices must overlap to ensure appropriate communication. Some repeaters and gateways have to be added and some sensors were moved when their position is not fixed in the room to ensure connectivity to the gateway.

3.1 Deployment Challenges

Several difficulties have been faced using this deployment approach. The installation of 433 MHz technology required the use of a custom application to visualize live connectivity information during the deployment. However, a second gateway has been used because the data collection on a unique point is not possible. For the technologies in the 2.4 GHz range, poor reception quality was experienced due to interferences even when the channels used by the different technologies are considered prior to deployment (e.g. Plugwise—Ch 15 Waspmote—Ch 12). Moreover, intermittent WiFi experiments are undertaken in this building leading to disruption on the connection. For the Plugwise technology, having 2 gateways would save 6 repeaters, and therefore more sensing nodes could be deployed as required. However, this extra gateway increases the total bill of material. It was found that the external thermal insulation of the building introduced some adverse effects on performance. For some technologies, signals from the 1st floor cannot reach the 2nd floor through open air because of the metallic grid on which the thermal insulation bricks are fixed. Also the restriction of having one unique collection point for all measurements has also led to an additional deployment complexity as repeaters have to be added. From a global cost perspective, and taking into account the time spent by the installer to manually tune the optimal position of these repeaters, a solution with multiple gateways might have been a better option. The effort and number of iterations required to get all the technologies working correctly can in some scenarios outweigh the savings made in terms of additional devices. Also, the manual tuning and troubleshooting of device/repeater positions by experts incurs additional salary cost.

4 Design Tool Deployment

There are a number of commercially available propagation modeling software tools but are largely focused on the design of IEEE802.11 (WiFi) based wireless networks. The tool that was used in this case was developed at Cork Institute of Technology [5]. The objective of the tool is to support the selection of the optimal configuration (number and position) of sensor, repeater and gateway devices in order to maximize link quality across all nodes while minimizing the overall infrastructure cost. A detailed description of the features of the design tool can be found in [5]. The design tool was used to create the deployment plans for all technologies. The user of the tool was only provided with the system requirements and a map of the building, the user was not on site prior to completing the designs.

Using the tool the user first defines the environment where the network should be deployed, the Auto-Cad file of building floors was imported into the tool and processed to allocate predefined material types for the walls, doors and windows. This is used along with the technology specification (frequency, transmit power, sensitivity threshold) as inputs into the tool. The tool has two options an automatic optimization, which will suggest the optimal position of gateways and repeaters to ensure reliable connectivity with predefined sensor positions. Alternatively, the user can place, move and manually adjust device positions and evaluate the expected link quality for the given configuration. Removing the constraint of a single gateway the tool suggests two gateway devices are required to ensure reliable connectivity. For this site, the designs with a single gateway were provided to the installer based on which they positioned the devices in the building.

5 Results Analysis

To present the benefits of using the design tool approach, Table 1 summarizes the savings that can be achieved. The environment preprocessing is required for the design tool as it is a critical input for estimating the coverage within the building. This involved drawing the walls, doors and windows and allocating material types to these objects. Regardless of the number of technologies that are considered in the environment this definition is only done once, this time can reduce/increase depending on the availability of the information for a specific site. This step is not necessary for the baseline as a representation of the floor plans is already available

Table 1 Results summary

Technology	Baseline			Design tool			Savings (%)
	Design time (min)	Deployment time (min)	Deployment iteration	Design time (min)	Deployment time (min)	Deployment iteration	
Environment preprocessing	–	–		60	–		0
Oregon/ Chacon	100	422	0	24	65	0	83
Homes	0	5	0	0	5	0	0
Waspmote	60	180	8	14	40	4	77
Plugwise	60	360	9	44	210	6	40
TelosB	25	60	0	14	60	1	13
Watteco	30	150	5	24	40	0	64
Total	275	1,177		180	420		59

in the appropriate format to be able to overlay coverage circles. For all technologies, significant savings can be achieved. This can be attributed to the ease of use of the tool and the ability to output an initial design that is close to the final installation. The deployment iteration represents the number of changes required to the original design in order to successfully get all sensors communicating with the gateway devices. In most cases this number is reduced by the design tool. Across all technologies a total of 59 % time savings can be accomplished which can be reflective of labor cost savings that can be achieved for this specific environment. To evaluate the performance differences with the design outputs a comparison between deployments for the TelosB technology is presented here. The TelosB application provides more fine grained information regarding the network therefore provides the capability to analyze the network performance.

Figure 2 presents the deployment plans as used by the installer. On the left of this figure is the baseline deployment map where three repeaters were required (only one was planned at design stage), two on the first floor and one on the second floor to connect the devices on the first floor to the gateway placed on the second floor, as three repeaters are required one sensing location is removed from the original planned deployment. On the right of this figure is the design tool output which implies that only two repeaters are required, one on the first floor and one on the second floor, therefore all sensor locations are satisfied.

To compare the quality of each deployment the first thing that was considered was the packet reception rate (PRR). The goal is to establish which network configuration is the most reliable in terms of the number of packets delivered versus the expected number of packets when assigned a fixed sensing interval.The PRR over the complete deployment period of one week showed that the baseline deployment performed significantly better than the design tool. Both designs ensure 100 % of packets are

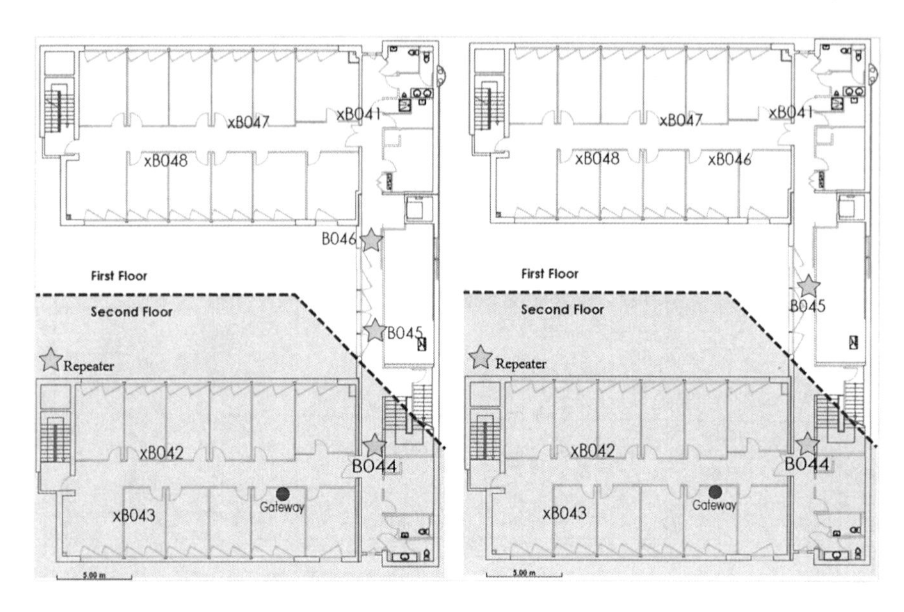

Fig. 2 Deployment plans. **a** Baseline. **b** Design tool

received from nodes deployed on the second floor which is expected as they are close to the gateway. For the design tool deployment the nodes on the first floor suffer around 30 % packet loss over the duration of the deployment which is a significant performance difference with only 5 % of packets dropped with the baseline configuration. Node B044, which is used to route data from nodes on the first floor to the gateway, is a single point of failure within the network. When the link between this node and the gateway is lost, all other nodes on the first floor show similar poor PRR. However, the location of node 0xB44 is the same in both baseline and design tool topologies. This would suggest either a problem with the nodes on the first floor reaching B044 in the design tool deployment or a change in the condition of the environment influencing the connection between B044 and the base station.

To further investigate the dependency on node B044, Fig. 3. shows the end-to-end PRR over the total duration of the deployment for nodes B044, B045 and B046. This

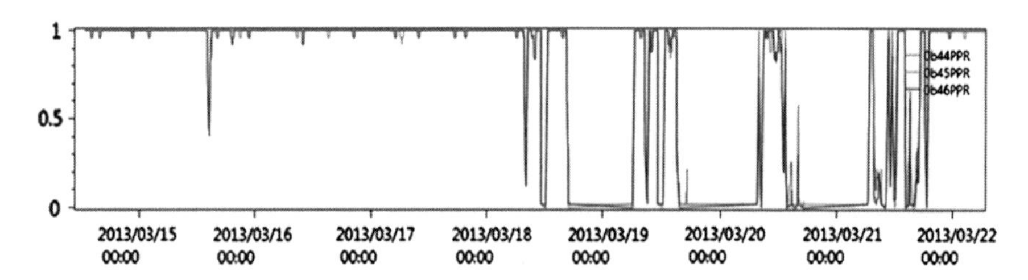

Fig. 3 The PRR of nodes B044, B045, B046 for design tool deployment

figure shows that the reception of the nodes during the first four days was almost perfect. It may be important to note that the 16th of March is a Saturday and the 17th is a Sunday. Suddenly, in the morning of Monday 18th of March, the link begins to fail; the packet reception rate drops to 0 % during the subsequent three nights and remains unstable during the rest of the days until the night of Friday the 22nd. The sudden poor link quality over, could be attributed to changes in the environment conditions e.g. the presence of people that may not have been occupied to the same extent over the weekend period. However the link starts to completely fail during the nighttime for long periods of time, and it then continues to have more sporadic connection loss during the day. As can be seen from the figure the same pattern is observed for the nodes connected to B044 verifying that the loss of packets from the nodes on the first floor is caused by the failure of the link from the node B044 to the gateway.

The PRR for the baseline deployment resulted in a similar pattern albeit not as severe as the design tool deployment. The overall reception rate is much higher than the design tool deployment but this is due to the good quality of the links during all nights. The majority of the total packet loss ratio (5 %) falls during daytime. In both deployments the link between B044 is the most problematic. The exact cause of this is difficulty to establish but it is suggested that a change in the environment configuration (e.g. fire door closed at night) during Design Tool deployment, which was not considered by both designs, may have been responsible for the link failures. As devices remained in place for 3 more days (during Easter weekend), additional data is also available for the baseline deployment. This shows a very similar pattern to that of the design tool deployment. All nodes from B044 only managed approximately 62 % reliability. Once again, this stresses the possibility of environmental factors influencing the performance of the link between B044 and the gateway and not necessarily a result of the differences between the design outputs. Received Signal Strength Indicator (RSSI) readings from nodes were used to analyze deeper the cause of failure. Based on previous experiments and as shown in [6], it can be determined that when the RSSI values falls below −90 dBm the link quality drastically decreases, and the packet reception rates will in turn decrease. Some links are close to the −90 dBm but on average the most used links remain above this threshold, however these links will be more sensitive to changes in the environment such as interference sources.

The most problematic link, the one from B044 to the gateway presents an average RSSI of −79 dBm for both designs. Theoretically, this would not indicate that the link failure was caused by a weak signal and also that, if minor changes occur within the environment, it should be able to absorb the signal degradation. When the RSSI values are analysed for the link between B044 and the gateway over the complete deployment period they appear reasonable. As the RSSI is only extracted from received data packets, the RSSI is unknown when there is no connection. However, there is no indication from the RSSI that a failure is about to occur. It is difficult to ascertain exactly the cause of the failure but it has beenshown that the link between B044 and the gateway is problematic, and also that the design tool deployment is most affected by this link failure. It is proposed this is due to changes in the environment but it is very difficult to verify this assumption.

6 Lessons Learned

The savings gained by the use of the design tool emphasize the need for support tools particularly for indoor environments. However the dynamics of the environment are very difficult to predict and can result in poor performance no matter how careful the design process has been, therefore the tools need to extend beyond design and more towards holistic management tools. The following are considerations that will be taken into account as future work [7].

- Link models only provide a "one-shot" prediction and cannot model the complex nature of the environment dynamics
- Tools should bridge the gap between operation and design to learn how to create better designs but also to actively adapt a configuration to handle different operating conditions
- A metric to indicate the dependency of nodes on each other should be provided to the installer from the design tool output
- The design tool should consider redundancy in its optimization process.

Acknowledgments The authors wish to acknowledge the support of the EU Commission under the FP7 SCUBA project (Grant Agreement No 288079) in part funding the work reported in this chapter.

References

1. Reinisch, C., Kastner, W., Neugschwandtner, G., Granzer, W.: Wireless technologies in home and building automation. In: 5th IEEE International Conference on Industrial In-formatics, vol. 1, pp. 93–98 (2007)
2. Itard, L., Meijer, F., Vrins, E., Hoiting, H.: Building Renovation and Modernisation in Eu-rope: State of the Art, review. Jan 2008.
3. Ringwald, M., Romer, K.: Deployment of sensor networks: Problems and passive inspec-tion. In: 5th Workshop on Intelligent Solutions in Embedded Systems (2007)
4. EnOcean Wireless Systems—Range Planning Guide http://www.enocean.com/fileadmin/redaktion/pdf/white_paper/WP_RANGE_PLANNING_Jun09_en.pdf
5. Mc Gibney, A., Guinard, A., Pesch, D.: Wi-design a modelling and optimization tool for wireless embedded systems in buildings. In: IEEE 36th Conference on Local Computer Networks (LCN) (2011)
6. Srinivasan, K., Levis, P.: RSSI is under appreciated, In: Third Workshop on Embedded Networked Sensors (2006)
7. FP7 SCUBA Project http://www.ict-scuba.eu

Long Term WSN Monitoring for Energy Efficiency in EU Cultural Heritage Buildings

Femi Aderohunmu, Domenico Balsamo, Giacomo Paci and Davide Brunelli

Abstract Historic buildings are a distinctive and invaluable characteristic of numerous European cities, and living symbols of Europe's rich cultural heritage. However, today, EU cultural heritage buildings are contributing huge percentage to the greenhouse gas emissions. This has led to the increasing of wireless sensor network (WSN) deployments aimed at monitoring and improving the energy efficiency of these historic buildings. In this chapter we show a long term, low cost, passive distributed environmental monitoring system that promotes energy-efficient retrofitting in historic buildings. We focus on the design and implementation of an innovative technological framework, and on the hardware and software development of the solution. The presented system provides real-time feedback for the civil engineers for prompt intervention via remote interfaces.

1 Introduction

Intelligent monitoring of the structural health of historic buildings has become an important research problem domain. Particularly, due to the lack of pre-existing mon-

F. Aderohunmu
Information Science Department, University of Otago, Dunedin, New Zealand
e-mail: femi.aderohunmu@otago.ac.nz

D. Balsamo · G. Paci
Department of Electrical, Electronic and Information Engineering (DEI),
University of Bologna, Bologna, Italy
e-mail: domenico.balsamo@unibo.it

G. Paci
e-mail: giacomo.paci@unibo.it

D. Brunelli (✉)
Department of Industrial Engineering (DII), University of Trento,
Via Sommarive 14, Trento, Italy
e-mail: davide.brunelli@unitn.it

K. Langendoen et al. (eds.), *Real-World Wireless Sensor Networks*,
Lecture Notes in Electrical Engineering 281, DOI: 10.1007/978-3-319-03071-5_25,

Fig. 1 **a** Heritage building with WSN deployment. **b** Hardware unit: A rear view of 32-bit WISPES W24th prototype

itoring models and to the constraints associated with the measurement devices. With the increasing ease of fast prototyping and low cost of manufacturing, miniaturized smart sensor nodes can be easily configured for monitoring purposes, today. The challenge is however, the ability to allow these nodes to operate in a collaborative manner over a long period of time without user intervention. The development of such architectures require a cross layer design from hardware and network perspective to the application layer. In this chapter we introduce an application scenario and we present a design testbed and the implementation strategies for a rapid, accurate and low-cost intelligent monitoring. The selected hardware allows low power consumption and long distance peer-to-peer data transmission. The software design is an open source built on the NXP Jennic platform [12]. It supports star, tree and linear network topologies. An integrated prototype has been deployed in a 36-nodes WSN testbed in a 4-storey renaissance building of about 1300 m^2, see Fig. 1a. The building is currently hosting several employees and students, and at the same time contains important historical artworks and frescoes. One of the main source of concerns for the local conservation board is how to preserve these artworks. It is necessary to conduct a timely estimation of any potential risk that could damage the building. This requires a real-time monitoring and an appropriate response model that can allow easy reproduction of the structural characteristics of the building.

The goal of the work is to develop a low-power, long-term and low-cost hardware and software solution that can efficiently monitor local climate changes in the building. Our design consists of WISPES 24th node type [20], which gathers environmental parameters such as, air temperature, relative humidity, vibration, gas concentration, air quality and ambient light. The network delivers data through a multi-hop network topology.

The remainder of this chapter is organized as follows: In Sect. 2 we present the application scenario, followed by the description of the hardware prototype in Sect. 3. In Sect. 4 we describe a proprietary software stack architecture that has been adapted specifically to meet the needs for low-cost, long-term and low-power operational life-cycle. In Sect. 5 we present our deployment testbed, it is worth mentioning that, the deployment has currently spanned 18 months and it is assessed to be in the half of the lifetime. Finally, in Sect. 6 we conclude the chapter.

2 Application Scenario

Structural health monitoring of building is a critical component of the civil infrastructure system. For instance, heritage buildings are often on the lookout for health monitoring by civil engineers due to their touristic attraction towards the public. The safety of such buildings is necessary to limit the danger it poses to the economic vitality of the society especially during damaging events such as earthquake etc. The conservation of such structures for the future generations is one of the main tasks of the European Union (EU). It is increasingly becoming important to understand the deterioration processes which affect artworks in museums and historic buildings, [9, 21]. Preserving the health of such structures requires a continuous monitoring process of the building [2]. Some of the parameters of interest are temperature, humidity, dust presence, quality of the air, etc. Apart from monitoring the health of the building structures, some of the heritage buildings contain some historical artifacts and frescoes that require a certain temperature and humidity levels for easy preservation.

The development in the fields of micro-electromechanical systems and wireless communications have introduced easy and low cost monitoring of such buildings, [1] and [13]. Therefore, it is possible to develop low-power, easy to install and miniaturized gadgets to sense and transmit data at a reasonable distance to more powerful gateway nodes equipped with high computational capabilities in place of the common commercial cabling system that is often utilized. Similar application design using WSN have been used in recent years. In Villa Regina [15], authors deployed a monitoring system composed of a wireless network of smart buttons, specifically designed to satisfy the requirements for applications in the field of cultural heritage. Majority of their work focused on evaluating the reliability of WSN for protecting and preventing the oxidation of old mirrors. Similarly, in the case study of Torre Aquila [7]; a medieval tower in Trento (Italy), where a customized hardware efficiently collects high-volume vibration data using specially designed sensors to acquire the building's deformation. One of the main drawbacks of these existing deployments is that, the majority of the implementations do not support ultralow power operation, thus hampering their use for a long-term deployment. Since the lifetime of a WSN depends largely on the power consumption at each sensor node, we focused on reducing power consumption both when the node is in active mode and when the node is in sleep mode. We are able to achieve this by optimizing the radio activity through a synchronous power management strategy and through a dynamic power control algorithm. This minimizes the use of the DC–DC converter, which guarantees a stable input voltage reference when the system is in 'ON' mode. In the next section, we describe the hardware unit for our monitoring application.

Table 1 Summary of the energy consumption for one cycle of the state machine

Sample rate	No transmission (mJ)	Transmission (mJ)
10 s	8.184	27.894
1 min	9.5	29.21
10 min	23.72	43.43

3 Sensing Unit and Hardware Design

The hardware unit is shown in Fig. 1b. The sensor node is built on the NXP JN5148 module with interconnected on-board sensors. The core of this module is designed to meet the requirements of ZigBee PRO applications, nevertheless we optimized the network protocol with proprietary features. The W24th nodes are equipped with the following on-board sensors: Temperature (0.01 °C resolution; −40 to +90 °C. range), Humidity (0.04 °C resolution; 0–100 % rh range), Light sensor (0.23 lx resolution; 0–100,000 lx range), 3-axis accelerometer (1 mg resolution; 0–6 g) and MOX Gas sensor interface (10 % resolution). Our prototype incorporates an integrated hardware and software designs that is capable of supporting both tree and star topologies. The flexibility of the hardware permits the W24th node to serve indifferently as a coordinator node, sleep-end node or router depending on the chosen topology. Through the wireless communication channel, the coordinator node can broadcast packets to the entire network, and the network can be organized. The sensing units are able to sense all the previously mentioned parameters, store the data, analyze and transmit the data when necessary Table 1.

The core of the system is a 32 bit micro-controller with a 2.4 GHz radio transceiver that can achieve about 100 m communication range in an indoor-environment. The 32 Mhz CPU clock permits the development of distributed data processing applications with the possibility of attaching an external memory slot i.e., W24th is equipped with a microSD card reader for local storage, data logging and backups. This allows the W24th to be connected to several expansion boards such as multi-channels acquisition interfaces, actuators, USB, KNX devices, energy harvesters and more [6, 10, 18, 19]. Often in the field of wireless sensor networks, wireless communication is known to be one of the major sources of power consumption. Long range communication is normally avoided as it increases the power drainage of the transceiver. In ambient monitoring application, communication range could be up to hundreds of meters. It is imperative to avoid constant transmission when not necessary to prolong the battery life. It is possible to achieve local data processing such as aggregation from cluster-heads or coordinator node due to the support of a peer-to-peer communication offered by nodes. The W24th uses an aggressive power management method using 8 μA in sleep mode, 17 mA in transmit mode and 23 mA in receive mode, which guarantees a longer battery lifetime. For example, compared with a TelosB mote [8], our node platform is able to yield about 35 % more power savings. The raw data transfer rate of the transceiver is up to 26 kps when used on the communication protocol and it includes reliable re-transmission/acknowledgment procedure according to the IEEE 802.15.4 standard.

Dynamic power-supply control In our test deployment one of the focus is to reduce the overall power consumption both when the node is in active mode and in the sleep mode. In the active mode, we achieved a reduced power consumption by optimizing the radio activity through a synchronous power management strategy. Similarly, in the sleep mode, reduced power consumption is achieved by implementing a dynamic power-supply control algorithm, which minimizes the use of the DC–DC converter.

We implement a novel technique for extending the lifetime of a WSN node, by minimizing the consumption of the DC–DC converter during sleep mode. This is accomplished through an adaptive algorithm that turns 'ON' and 'OFF' a DC–DC converter to minimize the on-board sleep power consumption. DC–DC converters, typically, need an output capacitor for high frequency ripple filtering. The adaptive algorithm computes on-line the maximum interval which guarantees the node to operate solely with output capacitor, taking into account both the total leakage and its variations under temperature changes. The proposed solution significantly increases the lifetime of applications with low-duty cycle activity, such as WSNs, achieving in the performed test a lifetime increment of over 30 %. In this way, the system allows a reliably data transfer and so far has an estimated lifetime beyond three years in full deployment.

4 Software and Protocol Stack Architecture

The protocol stack design for our intelligent monitoring application is built on the Jennic platform [12] developed on the IEEE 802.15.4 standard, which operates in the globally unlicensed 2.4 GHz band. Our network design consists of a Coordinator node which serves as the sink node (central server), some Router nodes and several Sleep-End nodes.

The Jennic platform ensures real-time data logging at the Coordinator node due to the memory management strategy in place. The design of the protocol stack has the following features: (1) Network load balancing to avoid data throughput congestion; (2) Offers support to the Sleeping End nodes for extended battery life; (3) Provides end-to-end message acknowledgment; (4) Allows over-the-air download; (5) Provides statistics for network maintenance (6) Supports automatic route formation and repair in-case of routing and (7) Finally, it provides network re-shaping to reduce network depth. In addition to these features, one of the main characteristic of the Jennic platform is that it can efficiently handle all the network traffic and can manage network faults with a self-healing mechanism. All these features make Jennic platform a choice for our WISPES W24th node prototype.

4.1 Time Synchronization

Time synchronization is a fundamental service in many distributed systems. Non-uniform operating temperature among deployed nodes and dynamic variation of the temperature is the major cause of clock drift in the nodes of a network. In this network protocol we implemented a low-overhead clock synchronization for WSNs based on a new temperature compensation algorithm (TCA) presented in [3]. The TCA is local and uses a temperature sensor to remove the effects of environmental temperature changes and to increase the time between synchronization intervals. Using the TCA, the 32 KHz unit achieves an effective clock drift of less than 5 ppm over a wide range of operating temperatures (−5 to 62 °C), a significant improvement with respect to the 55 ppm featured without temperature compensation.

This allows the coordinator node to broadcast beacons signals for all nodes in the network to synchronize their time-stamp, and then all the node including router can switch to the sleep mode, shutting down the radio. In the compensated node, the maximum drift is about 100 us, every 20 s, across the full temperature range; this means that a deviation in the network of 1 s may happen only after 55 h in any possible temperature condition. This synchronization accuracy is enough to put all the nodes of the networks (including routers) in sleep mode.

Moreover, the goal of our platform design is to be able to support other topologies, thus our design strategy deviates from the beacon-enabled approach. We instead use a non-beacon-based approach, which does not provide time synchronization and it does not support the active/sleep scheduling and Guarantee Time Slot (GTS) mechanism. It however, exhibits certain features, such as high scalability and low complexity that makes it a method of choice for our design. To be able to use the non-beacon-based approach, we need to re-design a method to synchronize the network. As a result, exploiting the temperature compensation algorithm (TCA), we implemented also an Energy-efficient Service-Packet-based (ESS) transmission scheduling algorithm. This method is based on a centralize approach that is capable of providing time synchronization on the top of the non-beacon-enabled IEEE 802.15.4 standard. The algorithm is designed to avoid collision in the network during packet transmission through a TDMA technique. The service-packet contains the time stamp and other necessary network information such as the active sensors on the network. The service-packet transmission across the network is achieved by using a novel method based on constructive interference of IEEE 802.15.4 symbols for fast network flooding and implicit time synchronization. The overall idea is that interference is considered an advantage rather than a problem. Based on this intuition, the simultaneous transmission of packet in the network becomes constructive. This method has proven to achieve a flooding reliability above 99 %, [11] which is quite close to the theoretical lower latency bound. Our method proves to achieve a network-wide time synchronization at no cost since it implicitly synchronizes the nodes as the flooding packets propagates through the network. With this approach, the nodes turn on their radios, listen for communications over the wireless medium, and transmit over-heard packets after a certain time delay. Moreover, since the neighbours of a

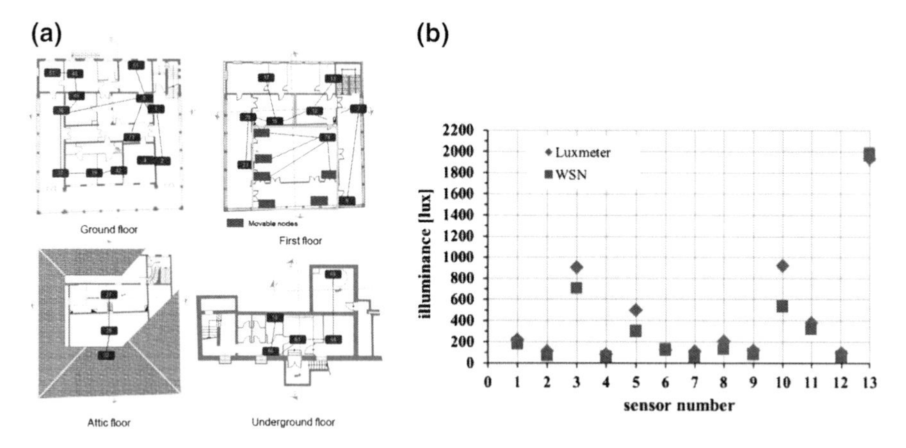

Fig. 2 Deployment map (**a**) comparison between illuminance measurements taken using both lux-meter and the deployed WSN (**b**)

sender recieves packet at the same time, they also transmit the packets at the same time.

One key strategy to ensure that a received packet has been relayed, is to allow each node decrement an internal counter for the number of Service Packet transmitted. To summarize, the network node achieves synchronization by using service packets and the coordinator defines the network topology for the deployment based on the available Link Quality and finally, it schedules each node to their given time-slot.

5 Deployment

Our deployment consists of 36-node setup located in the different parts of a 4-storey historic building shown in Fig. 2a. The reason for the different positions of the nodes is to be able to capture various characteristics of environmental variables across the building. For example, the first-floor contains ventilation systems that are constantly interrupted by the occupants; the attic contains generators and it is more closer to the roof (outside) temperature, while the underground floor has less interaction. Understanding the dynamics of the occupancy with respect to the characteristics of the building is of interest to the civil engineers we interacted with. In our deployment, 13 nodes are used as mobile stations for specific measurement purpose. These nodes can be used for dynamic surveys and may be employed to confirm the data collected by alternative testing techniques.

The huge amount of data generated by different sensors is handled using Compressive Sensing algorithms [4, 5]. We present here only the illuminance measurement inside one of the rooms by using both professional lux-meter and 13 WSN nodes (Fig. 2b). Notice that this comparison demonstrates how our nodes, equipped with light sensors, can be used in place of the professional lux-meter to achieve a continuously monitoring of the daily or seasonal variations of ambient light with minimal efforts and errors on the measurements.

6 Conclusion

In this chapter, we describe the design and the implementation of a Wireless Sensor Network tailored for monitoring heritage buildings. The continuous monitoring of environmental parameters is crucial for protecting cultural heritage and historical buildings. In this type of application, wireless sensor networks represent an alternative to traditional measuring instrumentations. The use of WSNs guarantees low-cost and long-term monitoring with a high rate of data extracted from the deployment for further analysis. The network will be enhanced with Smart Wireless Power Meters [16, 17] and Multimodal Video Sensor Nodes [14] as future works. The proposed sensor network includes functionalities for network management and efficient power supply management and provides real-time feedback for the civil engineers.

Acknowledgments The research leading to these results has received funding from the projects *3ENCULT* and *GreenDataNet*, both funded by the EU 7th Framework Programme. In addition, the authors would like to thank WISPES srl for the implementation of the prototypes

References

1. Abrardo, A., Balucanti, L., Belleschi, M., Carretti, C., Mecocci, A.: Health monitoring of architectural heritage: the case study of san gimignano. In: Environmental Energy and Structural Monitoring Systems (EESMS), 2010 IEEE Workshop on, pp. 98–102, 2010
2. Anastasi, G., Lo Re, G., Ortolani, M.: wsns for structural health monitoring of historical buildings. In: Human System Interactions, 2009. HSI '09. 2nd Conference on, pp. 574–579, 2009
3. Brunelli, D., Balsamo, D., Paci, G., Benini, L.: Temperature compensated time synchronisation in wireless sensor networks. Electron. Lett. **48**(16), 1026–1028 (2012)
4. Caione, C., Brunelli, D., Benini, L.: Distributed compressive sampling for lifetime optimization in dense wireless sensor networks. IEEE Trans. Industr. Inf. **8**(1), 30–40 (2012)
5. Caione, C., Brunelli, D., Benini, L.: Compressive sensing optimization for signal ensembles in wsns. Industrial Informatics, IEEE Transactions on (2013)
6. Carli, D., Brunelli, D., Benini, L., Ruggeri, M.: An effective multi-source energy harvester for low power applications. In: Design, Automation Test in Europe Conference Exhibition (DATE), pp. 1–6, 2011
7. Ceriotti, M., Mottola, L., Picco, G. P., Murphy, A. L., Guna, S., Corra, M., Pozzi, M., Zonta, D., Zanon, P.: Monitoring heritage buildings with wireless sensor networks: the torre aquila deployment. In: Proceedings of the 2009 International Conference on Information Processing in Sensor Networks, IPSN '09, pp. 277–288, Washington, DC, USA, 2009. IEEE Computer Society.
8. Crossbow. http://www.xbow.com (1995). Accessed 8 Sept 2012

9. Amato, F. D'., Gamba, P., Goldoni, E.: Monitoring heritage buildings and artworks with wireless sensor networks. In: Environmental Energy and Structural Monitoring Systems (EESMS), 2012 IEEE Workshop on, pp. 1–6, 2012
10. Dondi, D., Bertacchini, A., Larcher, L., Pavan, P., Brunelli, D., Benini, L.: A solar energy harvesting circuit for low power applications. In: Sustainable Energy Technologies, 2008. ICSET 2008. IEEE International Conference on, pp. 945–949, 2008
11. Ferrari, F., Zimmerling, M., Thiele, L., Saukh, O.: Efficient network flooding and time synchronization with glossy. In: IPSN, pp. 73–84, 2011
12. Jennic-Platform. NXP Jennic platform. http://www.jennic.com/index.php (1996)
13. Lynch, J. P.: Overview of wireless sensors for real-time health monitoring of civil structures. In: The 4th Intern. Workshop on Structural, Control, 10–11 June 2004
14. Magno, M., Brunelli, D., Zappi, P., Benini, L.: A solar-powered video sensor node for energy efficient multimodal surveillance. In: Digital System Design Architectures, Methods and Tools, 2008. DSD '08. 11th EUROMICRO Conference on, pp. 512–519, 2008
15. Neri, A.,et al.: Environmental monitoring of heritage buildings. In: Environmental, Energy, and Structural Monitoring Systems, 2009. EESMS 2009. IEEE Workshop on, pp. 93–97, 2009
16. Porcarelli, D., Balsamo, D., Brunelli, D., Paci, G.: Perpetual and low-cost power meter for monitoring residential and industrial appliances. In: Design, Automation Test in Europe Conference Exhibition (DATE), pp. 1155–1160, 2013
17. Porcarelli, D., Brunelli, D., Benini, L.: Clamp-and-measure forever: A mosfet-based circuit for energy harvesting and measurement targeted for power meters. In: Advances in Sensors and Interfaces (IWASI), 2013 5th IEEE International Workshop on, pp.200–205, 2013
18. Porcarelli, D., Brunelli, D., Magno, M., Benini, L.: A multi-harvester architecture with hybrid storage devices and smart capabilities for low power systems. In: Power Electronics, Electrical Drives, Automation and Motion (SPEEDAM), 2012 International Symposium on, pp. 946–951, 2012
19. Weddell, A.S., Magno, M., Merrett, G.V., Brunelli, D., Al-Hashimi, B.M., Benini, L.: A survey of multi-source energy harvesting systems. In: Design, Automation Test in Europe Conference Exhibition (DATE), pp. 905–908, 2013
20. WISPES. Wispes W24TH sensor node. http://www.wispes.com (2009)
21. Zonta, D., Pozzi, M., Zanon, P.: Managing the historical heritage using distributed technologies. Int. J. Architectural Heritage **2**(3), 200–225 (2008)

Printed by Publishers' Graphics LLC
CAMZ140209.15.15.133